Praise for *Slingshot*

Slingshot does for the launch loop what Arthur C. Clarke's *The Fountains of Paradise* or Sheffield's *Web Between the Worlds* did for the space elevator. Again, Williscroft delivers a great mix of hard science fiction and action.

—Alastair Mayer
Author of the *T-Space Series*

Robert Williscroft deftly crafts an energetic story around a phenomenal technological development just over the horizon: the space launch loop. The technical detail woven into this story is an education unto itself. But don't assume that Williscroft chooses raw infodump over story—*Slingshot* is an adventure that pulls you in, gives you characters that are engaging, and invites you to follow them through their challenges. What Williscroft has done in *Slingshot* is no easy task—he has balanced the hard aspect of science fiction with the character portrayals that those who despise that very hard science fiction beg for. The last decade has seen impressive leaps in the theoretical work toward the launch loop—this book couldn't come too soon! And you won't be able to keep from reading all the way to the end. Williscroft's art continues to be praise-worthy!

— Jason D. Batt
100 Year Starship
Author of *The Tales of Dreamside Series*

The engineering triumph portrayed in this book is staggering—an electromagnetic loop 80 kilometers high able to launch spacecraft safely into orbit without rocket fuel. A dedicated team of visionaries overcomes natural hazards and sinister opponents to complete this heroic project. The detail is convincing, and the action is entertaining.

John Knapman, PhD
Director of Research
International Space Elevator Consortium (ISEC)

Praise for *Slingshot*

I've been a fan of Robert Williscroft's books for a while now. They're action packed and filled with all kinds of interesting, real-world information. *Slingshot* fits right in.

Slingshot is about the development of an earth-bound spaceport in which spaceships are taken 80 kilometers above the Earth by elevator and hurled onto their trajectory by a very fast moving ribbon of soft iron. It is much easier, cheaper, and cleaner to launch spaceships from here due to the rarified atmosphere. This concept may be a reality someday. The book begins with a forward by Keith Lofstrom, the originator of this concept called the "launch loop."

Learning about the launch loop is the most interesting aspect of this novel. Williscroft's descriptions of the construction techniques, its operations, and the benefits for space travel are absolutely fascinating. The book takes place about thirty years in the future, and I could easily see such a project becoming a reality in that time.

The plot of the novel is driven by the development and construction of the project, which is being a threatened by ill-informed environmentalists bent on destroying the project. The launch loop is far greener than the current method of launching vehicles into space, but a sinister power has misled the environmentalists into believing that sabotaging the launch loop is saving the planet. Meanwhile, the sinister power is protecting its own economic interests.

As usual, Williscroft has a created a cast of interesting and driven characters. The book is a fascinating read, and you are guaranteed not only to learn a lot, but to dream about the future of space travel.

Marc Weitz
Past President
The Los Angeles Adventurers' Club

Praise for *Slingshot*

Real people, accurate science, and high adventure under the ocean, in the air, and in space. This is about the heroic men and women who will take us there, and the villains who try to stop them.

Keith Lofstrom
Inventor of the Launch Loop

Slingshot is a riveting, Hard Science Fiction novel about the construction of what will become the largest machine in all of human history. *Slingshot* is fiction, because it is describing something that has not happened…yet! Details of the construction, its theory and mathematical foundation, are all grounded in actual current science and engineering.

Conflict and suspense in the story are provided both by ecology terrorists who seek either to stop or destroy the "Launch Loop," as well as hidden adversaries plotting and manipulating in the background. Meanwhile readers follow the well-developed characters and gain insights into their wants, needs, and various sexual tensions that paint a fascinating backdrop to the story, while providing real insight into their motivations as the story takes increasingly interesting twists and turns.

James T. (Stew) Stewart
Author of *The King of a Thousand Suns*

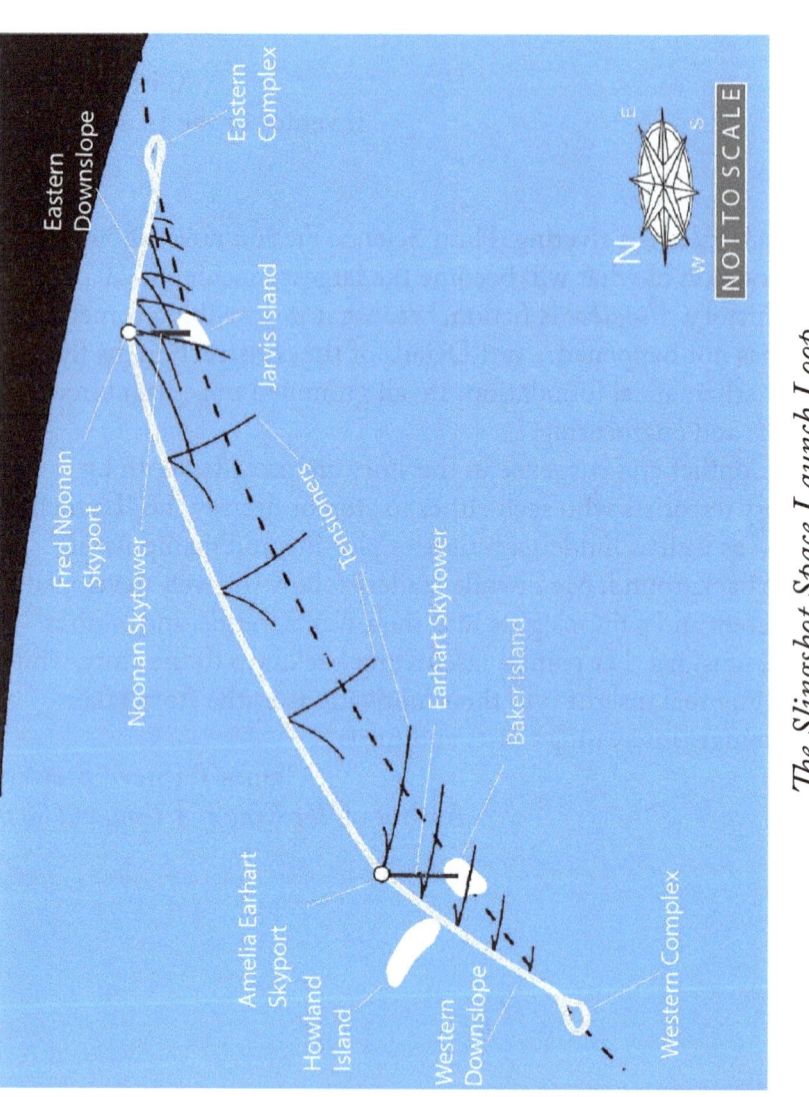

The Slingshot Space Launch Loop

VOL 1 IN THE STARCHILD SAGA

SLINGSHOT
— A NOVEL —
BUILDING THE LARGEST MACHINE IN HUMAN HISTORY

Robert G. Williscroft

Fresh Ink Group
Guntersville

Slingshot
Building the Largest Machine in Human History

Copyright © 2023
by Robert G. Williscroft
All rights reserved

Fresh Ink Group
An Imprint of:
The Fresh Ink Group, LLC
1021 Blount Avenue #931
Guntersville, AL 35976
Email: info@FreshInkGroup.com
FreshInkGroup.com

Edition 4.0 2023

Cover design by Gary McCluskey
Illustrations by Robert G. Williscroft

Except as permitted under the U.S. Copyright Act of 1976 and except for brief quotations in critical reviews or articles, no portion of this book's content may be stored in any medium, transmitted in any form, used in whole or part, or sourced for derivative works such as videos, television, and motion pictures, without prior written permission from the publisher.

Cataloging-in-Publication Recommendations:
SCI098020 SCIENCE / Space Science / Space Exploration
FIC028130 FICTION / Science Fiction / Space Exploration
FIC027130 FICTION / Romance / Science Fiction

Library of Congress Control Number: 2023911575

ISBN-13: 978-1-958922-28-6 Paperback
ISBN-13: 978-1-958922-29-3 Hardcover
ISBN-13: 978-1-958922-30-9 Ebook

Table of Contents

Praise for *Slingshot*
    Frontispiece—The Slingshot Space Launch Loop
Title Page
Table of Contents i
Acknowledgements vii
Foreword to the 1st Edition viii
Foreword to the 2nd Edition x
Foreword to the 3rd Edition x
Dedication xi
Cast of Characters xii
   Prologue 1
      IMAGE 1—Amelia Earhart in the cockpit
        of her Lockheed Electra 1
      Lockheed Electra—Above the Western Equatorial Pacific
   Part One—*He shall have dominion over the fish in the sea* 7
      IMAGE 2—The Slingshot Space Launch Loop 7
Chapter One 9
   Equatorial pacific—southeast of Baker Island 9
Chapter Two 18
      IMAGE 3—Diagram of the Western Complex 18
   Western Complex—300 km west of Baker Island 18
Chapter Three 24
      IMAGE 4—Baker Island 24
   Baker Island—Margo Jackson's Quarters 24
   Baker Island—Airstrip 30
Chapter Four 31
   Airborne to Western Complex 31
   Western Complex—300 km west of Baker Island 33
Chapter Five 38
   Eastern Complex—Circular Deflector 38
Chapter Six 47
   Underway between Baker and Jarvis Islands 47
   Buoy 1528 48
   On the surface at Buoy 1528 50

Table of Contents

Chapter Seven 55
- Submerged on gills at Buoy 1528 55
- On the surface at buoy 1528 59
- Airborne to Jarvis Island 63

Chapter Eight 65
- *IMAGE 5—Jarvis Island* 65
- On the surface at buoy 1528 65
- Aboard *Green Avenger* underway for Jarvis Island ... 66
- Jarvis Island recompression complex 67
- *IMAGE 6—Diagram of the Eastern Complex* 72
- Eastern Complex 300 km east of Jarvis Island 73
- *IMAGE 7—Cutaway of the tube* 77
- Jarvis Island recompression complex 80

Chapter Nine 82
- Airborne above the Eastern Complex 82
- Jarvis Island recompression complex 84

Chapter Ten 93
- Eastern Complex 300 km east of Jarvis Island 93
- Jarvis Island Compound 95
- Jarvis Island wharf 97

Chapter Eleven 100
- Jarvis Island Compound 100
- Jarvis Island tarmac 102
- Jarvis Island Compound 104
- Jarvis Island—Southern Beach 111

Chapter Twelve 114
- Jarvis Island Compound 114
- Eastern Complex 116
- Jarvis Island 118
- Jarvis Island Southern Beach 120
- *IMAGE 8—Wreck of the Barquentine Amaranth* 122

Chapter Thirteen 125
- Seattle—Smith Tower 125
- Jarvis Island Compound 129
- Seattle—Smith Tower 131
- Jersey City—New Jersey 135

Table of Contents

Chapter Fourteen 136
   Jersey City—New Jersey.. 136
   Seattle—Smith Tower 138
   American Samoa—South Pacific 139
   Seattle—Smith Tower 141
   Baker Island—Margo's quarters 142
   American Samoa—South Pacific 143
   Baker Island—Margo's quarters 144
Part Two...*and over the birds of the air*........................ 147
     IMAGE 9—*Amelia Earhart Skyport Illustration* 147
  Chapter Fifteen 149
   Equatorial Pacific—Aboard *Aku Aku* south of
     Western Complex... 149
   Equatorial Pacific—Submerged aboard *Alvin* 151
   Baker Island—Operations Compound................. 155
  Chapter Sixteen 160
   Equatorial Pacific—Submerged aboard *Wampus* 160
   Equatorial Pacific—Submerged aboard *Alvin* 165
   Equatorial Pacific—Aboard *Aku Aku*................... 167
   Equatorial Pacific—Overboard south of
     Western Complex... 169
   Equatorial Pacific—Aboard *Skimmer One* 169
  Chapter Seventeen 171
   Equatorial Pacific—Submerged aboard *Wampus* 171
   Equatorial Pacific—At W-1 and W-3.................. 174
   Equatorial Pacific—At W-1 175
  Chapter Eighteen 179
   Baker Island—Operations Compound................. 179
   Seattle—Smith Tower 180
   American Samoa—Pago Pago............................. 180
   Baker Island—Operations Compound................. 181
   Western Complex—Control Center 186
   Seattle—Smith Tower 186
   Baker Island—Operations Compound................. 187
  Chapter Nineteen 188
   Seattle—Smith Tower 188
   Howland and Baker Islands................................. 189

Table of Contents

Baker Island—Operations Compound 191
Chapter Twenty 198
   Baker Island—Operations Compound 198
   Amelia Earhart Skyport 203
Chapter Twenty-one 209
   Baker Island—Socket Compound 209
   Amelia Earhart Skyport 211
   Baker Island—Baker Socket 212
   Howland Island—Operations Center 214
   Amelia Earhart Skyport 215
   Baker Island—Baker Socket 218
Chapter Twenty-two 222
   Amelia Earhart Skyport 222
   Baker Island—Baker Compound 227
   Amelia Earhart Skyport 228
   Baker Island—Baker Compound 230
Chapter Twenty-three 231
   Baker Island—Baker Compound Workshop 231
   Amelia Earhart Skyport 234
Chapter Twenty-four 239
   Baker Island—Baker Socket 239
   Amelia Earhart Skyport 239
   Baker Island—Baker Socket 241
   Baker Island—Meyerton Landing 243
   Baker Island—Aboard Skimmer One 243
   Amelia Earhart Skyport 245
   Baker Island—Baker Socket 245
   Aboard *Skimmer One* between Baker and
      Howland Islands 245
Chapter Twenty-five 247
   Amelia Earhart Skyport 247
   Baker Island—Baker Socket 248
   Aboard Skimmer One between Baker and
      Howland Islands 250
   Baker Island—Baker Compound .... `252
   Baker Island—Eastern Beach 253
   Baker Island—Baker Compound 255

Table of Contents

Chapter Twenty-six 257
    Seattle—Smith Tower 257
    Pyongyang– DPRK (North Korea) 260
    Seattle—Airborne to Boeing Field 261
    Baker Island 262
Chapter Twenty-seven 266
    Seattle—Smith Tower 266
    Baker Island—Baker Compound 268
    Jarvis Island—Jarvis Compound 270
Chapter Twenty-eight 274
    Fred Noonan Skyport 274
    Jarvis Island—Jarvis Compound 279
    Fred Noonan Skyport 280
Chapter Twenty-nine 283
    Seattle—Smith Tower 283
    Jarvis Island—Jarvis Compound 284
    Baker Island—Baker Compound 286
    Jarvis Island—Jarvis Socket 288
    American Samoa—Pago Pago 289
    Baker Island—Operations Center 290
    Amelia Erheart Skyport 292
Chapter Thirty 294
    Seattle—Smith Tower 294
    Honolulu—Airport Marriott 295
    Baker Island 297
    Baker Island—Control Center 298
    Amelia Earhart Skyport 301
Chapter Thirty-one 306
    Baker Island—Control Center 306
    Baker Island—Underway on *Skimmer Three* 310
    Baker Island—Control Center 313
Chapter Thirty-two 314
    Seattle—Downtown 314
    Seattle—Smith Tower 318
    Baker Island 319
    Amelia Earhart Skyport 321
    Pyongyang– DPRK (North Korea) 325

Table of Contents

Afterword 326
   Near Baker Island—Submerged on *Wampus* 326
   Near Baker Island—Aboard *RV Amelia E* 327
   Amelia Earhart Skyport 327
Post a Review 329
Excerpt from the first chapter of *The Starchild Compact* 320
   Cassini II in the asteroid belt 320
   L-4—Mirs complex, three weeks earlier 320
   Praise for *The Starchild Compact* 335
Hyperchess Rules 338
About Robert G. Williscroft 342
Other books by Robert G. Williscroft 343
Connect with Robert G. Williscroft 344
Slingshot Glossary 345

Acknowledgments

Several people contributed to the creation of this book.

Most significantly, my wonderful wife, Jill, whom I first met when I returned from a year at the South Pole conducting atmospheric research, and who finally consented to marry me nearly thirty years later, pored over each chapter with her discerning engineer's eye. She kept my timeline honest, and made sure that regular readers could understand fully the arcane details of the Launch Loop and its linear drivers powered by OTEC generators.

Jill's daughter, Selena, and twin sons, Arthur and Robert, also read the manuscript, and provided their insights.

Keith Lofstrom, the inventor of the Launch Loop, went through each chapter checking my math and engineering. He made several observations and corrections that improved this book.

John Knapman, Director of Research, International Space Elevator Consortium (ISEC), Keith's colleague and fellow space settlement enthusiast, read the manuscript and pointed out several errors that I had overlooked, and made several excellent suggestions.

Hard science fiction author Alastair Mayer reviewed the manuscript and offered his scientific, engineering, and editorial insight.

Attorney and world-class Adventurer Marc Weitz identified a couple of problems that I had missed, to the betterment of the book.

John McCracken went through the manuscript with his eye for detail. The result is a better story.

A tip of the hat to Gary McCluskey for turning the cover from a sketch and several ideas into the breathtaking scene that graces the front of this book.

It goes without saying that any remaining omissions, errors, and mistakes fall directly on my shoulders.

Robert G. Williscroft, PhD
Centennial, Colorado
July 2018

Foreword to First Edition

Launching to space is not just a matter of getting out of the atmosphere. To stay up, you need to get into orbit, which means traveling at 7.7 kilometers per second or more (about 17,300 mph), starting at the equator. Today, we do this with rockets, but they are expensive and use most of their fuel just lifting the rest of the fuel. Very little of the fuel actually lifts the payload. Many skilled people are needed to build and launch rockets. Their salaries, divided by low launch rates, are why rockets cost so much. Is there another way?

Arthur C. Clarke's 1979 science fiction novel *The Fountains of Paradise* (inspired by Jerome Pearson's 1975 *Acta Astronautica* paper) introduced us to the space elevator, a 100 thousand kilometer superstrong cable extending vertically from the equator far into space, supported by centrifugal acceleration. Step into an elevator car on the ground, and step out in orbit a few days later. Inspiring! Mind opening! And still not achieved decades later, mainly because we don't know how to make material that is strong enough. Some of us said, "THAT won't work. But what if…?" Two years and a dozen what-ifs later, I discovered the launch loop, which will throw payload into orbit, not lift it.

Imagine a stream of water out of a fire hose. Without air friction, the stream might make a parabolic arc 20 meters high. Faster, and the arc goes higher and farther. A stream moving 7.3 kilometers per second would come down on the other side of the planet, and a stream moving 11 kilometers per second would keep going into interplanetary space. Wrap the stream in a frictionless hose, and… THAT won't work either. But what if…?

The launch loop: replace the water with flexible iron pipe, 5 centimeters outer diameter, 3 metric tons per kilometer, moving at 14 kilometers per second. Bend it to the curvature of the earth with a stationary magnetic track, 7 metric tons per kilometer, 2,000 kilometers long, at 80 kilometers altitude. Stabilize the moving iron with electromagnets controlled by fast electronics. Turn it around at the ends with powerful magnets, and complete the loop.

On the eastbound section, 5-metric-ton payloads ride on magnets designed for high drag, which accelerates payloads at 3 gees. Payloads exit the east end of the track between 7.7 and 11 kilometers per second, to equatorial low earth orbits, to the moon, or to interplanetary space. Launching a payload weighing 5 metric tons to low earth orbit consumes 180 megawatt-hours, about $15,000 worth of electricity, or $3 per kilogram of payload. Passengers will still need vehicles and air, but freight can be launched on wooden shipping pallets. This small launch loop can launch 2,000 5-metric-ton payloads to orbit per day. Heavier launch loops can launch thousands of standard 30-metric-ton intermodal shipping containers per hour. They can also store peak power for the global electrical grid. Space travel can be as cheap as ocean cargo travel.

The whole world launches 400 metric tons to orbit per year. Launch loops will be assembly lines to launch megatons of payload. They will open up new markets and business opportunities because of the low cost and routine nature of the operations. One example, my current focus, is Server Sky (*http://server-sky.com*), which will use space solar power to do the world's data center computing.

In the early 1980s, there were other *Fountains* what-if-ers. I published in an *American Astronautical Society Newsletter* and other journals, and presented at many conferences. Much of that is preserved at *http://launchloop.com*. In England, physicist Paul Birch wrote about orbital rings in the *Journal of the British Interplanetary Society*. Ken Brakke, a math professor in Pennsylvania, published his version of orbital rings.

Ken, Paul, and I met at one of the Space Studies Institute conferences at Princeton. We spent three days developing nomenclature, doing math, finding errors and fixes. I met Robert Williscroft in the mid-1990s in Philadelphia while I was on a trip to the East Coast. He had contacted me about a novel he was outlining—*Slingshot*. We spent a day together becoming acquainted and have kept in touch since then. *Slingshot* in its current form is a result of our brainstorming during that visit.

For years afterwards, Paul and I swapped ideas. Under Jerome Pearson's leadership, and with our friend John Knapman (*http://space-cable.org.uk*), we submitted grant proposals until Paul's untimely passing in 2012. We were friends, never competitors, though Paul was much better at reciting Tennyson. I hope one of us will be the first launch loop astronaut. Conflict makes great stories, but friendship makes great lives, so I now pass page control to my friend R. G.

Keith Lofstrom
Inventor of the Launch Loop

Foreword To The Second Edition

Once I completed *The Iapetus Federation*, the third book in *The Starchild Trilogy*, it was clear that I needed to make some adjustments to *Slingshot*. I had set up the chapters and sub-chapters in both *The Starchild Compact* and *The Iapetus Federation* to make it easier to follow the timeline of the story. So, I modified the chapter and sub-chapter titles in *Slingshot* to be consistent with this.

In this second edition, each chapter and sub-chapter has a title that indicates the location where the action is taking place. I also created a complete Table of Contents that reflects these titles. I trust these changes will make this book even more enjoyable.

Robert G. Williscroft, PhD
Centennial, Colorado
July 2018

Foreword To The Third Edition

This third edition contains some minor line edits, formatting changes, and some changes to the Glossary. I also shortened my bio and added several more books that I have written since the original publication of *Slingshot*.

Robert G. Williscroft, PhD
Centennial, Colorado
November 2021

Foreword To The Fourth Full-color Edition

This full-color fourth edition was puplished by Fresh Ink Group as the first book in *The Starchild Saga*.

Fresh Ink Group
Guntersville, Alabama
June 2023

Dedication

This book is dedicated to my son Jason, who may actually live to see the first Launch Loop begin to move people and cargo into space, and to the Mohawk High-Iron Men who will certainly play a major role in making it happen.

Cast of Characters

Launch Loop International (LLI)

Klaus Blumenfeld—German Field Engineer in charge of OTEC power; BS in physics from Darmstadt Technische Hochschule; MS and PhD in electrical engineering and power systems from Cal Tech.

John Boyles—Board Member LLI; shrewd attorney; arranged a quick hearing by the Supreme Court that opened the way for Slingshot to go forward.

Mable Fitzwinters—Board Chairman LLI; heavy-set; models herself after Dixy Lee Ray.

Margo Jackson—Field Engineer in charge of underwater construction; BS in physics from Berkley; MS from Duke University in Marine Engineering.

Rex Johnson—Board Member LLI; knows his way around the halls of Congress; acquainted with Carleton Montague and Delmer Woodward.

Alexander Regent (Alex)—Field Engineer in charge of project; BS in engineering and a double MS in aquatic and aerospace engineering from MIT; MIT Hyperchess champion.

Eastern Complex

Kelly Bjork (BJ)—Diver on Eastern Complex; from Minnesota; ex-Navy diver.

Calvin Gofort (Gofer)—Dive Supervisor on Eastern Complex; professional diver from English oil fields; Scottish accent.

Cody Hayden (Cody)—Lead engineer on Eastern Complex; Colorado School of Mines undergraduate; MIT PhD in Ocean Engineering; Senior engineer for the OTEC development team at MIT; several major water construction projects, including Lincoln II connecting Manhattan and New Jersey.

Max Heister (Maxie)—Diver on Eastern Complex; from New Jersey; former oil rig diver.

Sean Isaacs (Sean)—Diver on Eastern Complex; Boston Irishman.

Cast of Characters

Fred Jones (Jonesy)—Diver on Eastern Complex; from Philadelphia; strong Philly accent; former hard-hat construction diver; newest member of the dive team

Samantha McNabb (Sam)—Diver on Eastern Complex; from California; wreck diver (with Woody); pioneered rebreathers for recreation (with Woody).

Thomas Mickelson (Micky)—Diver on Eastern Complex; Nebraska farm boy; ex-Navy diver.

Jessie Oberlin (Jess)—Diver on Eastern Complex; from Philadelphia; German background, but speaks only English; ex-Navy diver.

Antonio Park (Tony)—Crane operator Eastern Complex; from Brooklyn; operated container cranes up and down the east coast.

Joe Paulson (Joey)—Diver on Eastern Complex; from Washington state; ex-NOAA diver.

Todd Pelton (Todd)—Diver on Eastern Complex; from Texas; former oil rig diver.

Douglas Ready (Doug)—Crane operator Eastern Complex; from Chicago; operated container cranes and general cargo cranes on the Great Lakes.

Carlos Sanchez (Tex)—Lead pilot with Eastern Complex; from Texas.

Roxford Stewart (Stu)—Lead diver on Eastern Complex—in charge of diving; Duke Univ Mech Eng; former NOAA Diver in charge of U.S. Government East Coast diving; joined GlobeTech (international oil rig contractor) as their senior underwater engineer/chief diver.

Rodney Weis (Rod)—Diver on Eastern Complex; from Georgia; black; ex-Navy SEAL.

Edward Wood (Woody)—Diver on Eastern Complex; from California; wreck diver (with Sam); pioneered rebreathers for recreation (with Sam).

Western Complex

Emmett Bihm (Bimmy)—Lead diver on Western Complex—in charge of diving; from Minneapolis; ex-Navy salvage diver.

Jefferson Carver (Jeff)—*Skimmer Two* skipper; from Bermuda; strong Bermuda lilt; blue-black skin.
Abel Kilker (Abe)—Diver on Western Complex; from Arkansas.
Apryl Searson (Apryl)—Diver on Western Complex; from Oregon; trained EMT.
Domingo Solak (Solak)—Diver on Western Complex; from Balkans.
Pearl Wells (Pearl)—Dive Supervisor on Western Complex; black; tough-as-they-come Gullah-Geechee oil rig diver from the Carolina barrier islands.
Noemi Wien (Noe)—Lead pilot with Western Complex; joined when Radcliff & Woodward supplied a third floater.

Amelia Earhart Skyport

Shift One

Lance Fairbank (Lance) –High-Iron worker; New Jersey; has worked on "high-iron" projects all over the world.
Peter LaFleur (Pete)—High-Iron Mohawk; lives in Brooklyn; speaks a few Mohawk words; traces his ancestry back to Kahnawak Mohawks, early 1900s; has worked on "high-iron" projects all over the world; great-great…grandson of Peter LaFleur, high-iron worker and photographer of mid-20th C.
Mike Swamp (Mike)—Shift One Boss; High-Iron Mohawk; lives in Brooklyn; fluent Mohawk speaker; traces his ancestry back to Kahnawak Mohawks, early 1900s; has worked on "high-iron" projects all over the world.

Shift Two

Thomas Daillebaust (Tom)—Shift Two Boss; High-Iron Mohawk; lives in Brooklyn; fluent Mohawk speaker; worked on Geo. Washington Bridge in N.Y., and on "high-iron" projects all over the world; great…grandfather is Louis Joseph Dybo, a.k.a. "Chief Great Fire," son of Thomas Joseph Daillebaust.
Mathew Munns (Matt)—High-Iron worker; from Texas; has worked on "high-iron" projects all over the world; Seidell's best friend.
Guy Roth (Guy)—High-Iron worker; from Chicago; has worked on "high-iron" projects all over the world.

Kelly Seidell (Kelly)—High-Iron worker; from Texas; has worked on "high-iron" projects all over the world; Munns' best friend.

Shift Three

Rodney Chrietzberg (Critz)—Shift Three Boss; High-Iron Mohawk; lives in Brooklyn; fluent Mohawk speaker; has worked on "high-iron" projects all over the world; great…grandfather David Chrietzberg broke back in 1980s.

Rudolph Pigman (Pigman)—"Capsule Attendant," later, part of Shift Three; from Boise, Idaho.

Fred Noonan Skyport

Shift One

Ajay Dybo (AJ)—Shift One Boss; High-Iron Mohawk; lives in Brooklyn; fluent Mohawk speaker; worked on Geo. Washington Bridge in N.Y. with Mike Swamp, and has worked on "high-iron" projects all over the world; direct descendant of Louis Joseph Dybo, a.k.a. "Chief Great Fire."

Prajeet Kahnawak (Jeeter)—High-Iron Mohawk; lives in Brooklyn; speaks a few Mohawk words; has worked on "high-iron" projects all over the world; traces his ancestry back to Kahnawak Mohawks, early 1900s.

Erik Maffet (Erik)—High-Iron worker; from Chicago; has worked on "high-iron" projects all over the world.

Miscellaneous LLI Personnel

Chad Mickey (Chad)—Chinook pilot.
Noah—Maori Chief Steward at Baker Complex.
Jake—Maori Chief Steward at Jarvis Complex.

News Crew

Loraine Kutcher (Lori)—Fox Syndicate Reporter; from Seattle.
Dexter Lao (Dex)—Fox Syndicate Holo-cameraman; Asian background.
Mary Martain (Mary)—Reporter for *Me TOO!* magazine.

Environment Inc. (EI)

Dane Curvin—*EI* Personnel; Yale general studies, classmate of Gotch; Yale Law School.

Darius Gotch—*EI* Site sabotage; Yale general studies, classmate of Curvin.

Katelynn Leete—*EI* Manufacturing sabotage.

Quinton Radler—*EI* CEO; Berkeley Computer Sciences; Harvard MBA.

Green Force

Carmine Endsley (Carey)—*Green Force* member; age 25; scion of a Silicon Valley family; UC Berkeley computer technology.

Tiffany Montague (Tiffany)—*Green Force* member; age 18; privileged, pampered daughter of old-money family with oil-based wealth; on sabbatical from Bryn Mawr, English major.

Carmina Pebsworth (Carmina)—*Green Force* member; age 19; wayward daughter of the San Francisco Pebsworth magazine publishing family; attending Berkeley, Journalism.

Bobby Pfaff (Bobby)—*Green Force* member; age 18; youngest of three children of a Connecticut ship building magnate; Yale dropout.

Lars Watson (Lars) [Bob Weingard] [Ed Flannigan] [Neil Lansing]—Leader of *Green Force*; militant environmentalist; "professional" student, starting with general studies in Evergreen State College in Washington state, and finally completing a dual Master's Degree in Environmental Studies and the History of Science from University of Washington.

Francesca Woodward (Francesca)—*Green Force* member; age 17; debutant daughter of the Delaware Woodward banking family; attending Radcliff on strength of Daddy's endowment; English Lit major.

Bruce Yoon (Bruce)—*Green Force* member; age 18; son of a wealthy Chinese clothing importer in Los Angeles; attending UCLA, journalism.

Incidental personnel
Lance Banes—Fake Deputy Federal Marshal.
Ronald Botts—Deputy Federal Marshal.
James Franks—Deputy Federal Marshal.
Allan Hutshell—Fake Deputy Federal Marshal.
Judge Kalolo Leaupepe—Federal Judge in Pago Pago; attended law school with John Boyles
Carleton Montague—Tiffany's father; major player in the oil industry.
Mindi Pebsworth—Carmina's mother; married to Reyes Pebsworth, owner of a San Francisco magazine conglomerate.
Delmer Woodward—Francesca's father; heads the Delaware Woodward Banking interests.
George Yoon—Bruce's father; runs a large Los Angeles clothing import firm.

Democratic People's Republic of Korea (DPRK—North Korea)
General Jon Yong-nam—Aging advisor to Kim Jong-un; started out as a friend and advisor to the newly-minted Kim Jong-un; secretly controls DPRK Intelligence Operations including web-hacking efforts; controls and markets DPRK's missile & launch rocket exports.

VOL 1 IN THE STARCHILD SAGA

SLINGSHOT
—— *A NOVEL* ——
BUILDING THE LARGEST MACHINE IN HUMAN HISTORY

Slingshot

Image 1—Amelia Earhart in the cockpit of her Lockheed Electra

PROLOGUE

LOCKHEED ELECTRA—ABOVE THE WESTERN EQUATORIAL PACIFIC

"I'm tired, Fred. How much farther to Howland?"

She peered out through the Lockheed windscreen at the endless expanse of Pacific Ocean in front of her.

"Three hundred miles, Doll, just three hundred miles more on this leg. How're you doing up there?"

The lanky, soft-spoken man looked over his left shoulder at the dungaree-clad woman grasping the control wheel in front of her. In response, she rubbed her hand across her forehead and squinted into the reflected glare from the ocean surface far below. She glanced at her watch and then at the array of instruments in front of her.

"Hundred fifty knots, Fred. How do you make the fuel?"

Fred manipulated the circular dials of his navigator's slide rule. "Fine, Girl. We got more than enough."

He twisted around and peered into the periscopic sextant mounted in the cabin overhead. After jotting down a few numbers, he noted the time and checked a volume on the small table jutting out from the bulkhead in front of him. Then he turned, scowling, and took a second sighting of the sinking sun behind them. A few moments later he laid aside his reference book and said, "Drop down a thousand feet, will you? We seem to be bucking a pretty strong headwind up here."

The silver bird dipped its nose in response. The altimeter needle spun until it pointed to 11,000 feet. With her right hand, the pilot picked up a pair of binoculars and scanned the horizon in front of her. Fred took another sight on the sun and plotted his results. She turned around and looked at him expectantly.

"We're goin' the right way, Doll. But these running fixes—you know the assumptions you have to make...drop another thousand feet, will you please?"

Blood red water astern swallowed the sinking sun as inky blackness spread across the sky before them. She had planned it this way, one last evening star fix to establish more accurately their position before setting a final vector for Howland Island just a few miles north and east of where Equator and International Date Line cross.

July 2, 1937, was drawing to a close; but it would start all over again as soon as she crossed the 180th meridian. Outside, twilight quickly deepened to tropical night. A million stars twinkled to her left and overhead in patterns long familiar, while she had to will the bright points off to her right into recognizable patterns.

"Amelia, Doll," Fred turned to look at the pretty pilot wearing a leather skullcap, flaps dangling near her chin. "We got a problem, Girl."

She looked at him with a steady gaze, saying nothing.

"That head wind," Fred paused to push a pencil back to the center of his table. "It was a good deal stronger than we estimated. There's still a lot of water ahead of us, Doll, a lot."

She raised an eyebrow.

"Marginal," he said. "The fuel's marginal."

She pursed her lips and scanned the horizon once again in the growing darkness.

"You won't find it," he sighed. "It's still four hundred fifty damn miles out there."

On they flew into the darkness—one hour, two hours. On Fred's chart their plot line crept closer to a dot labeled Howland Island, but that was countered by the fuel gauge needle creeping toward empty.

Fred took another sighting. This time his fix seemed a bit farther from the last than it had from the preceding one. He turned and said into the darkness, "I think we picked up a westerly. Looks like our luck is holding." Then he scanned the horizon before them through his binoculars. "But no sign of the *Itasca*," he said and began fiddling with his transmitter dial.

"KHAQQ calling *Itasca*. We must be on you but cannot see you…gas is low…."

Static was the only response.

"Damn fool instrument!" Fred snarled, as he tried to zero into the homing signal he knew was being transmitted by the Coast Guard Cutter. "I guess we should have brought that new-fangled high-frequency receiver after all. It really didn't weigh that much." He grinned at Amelia's silhouette. "Good thing for the westerly, Girl. We need a kick in the rear!"

"A westerly," the woman's voice echoed from the pilot's seat. In her mind she could see the warning printed near the chart margin: *NOTE 3—Surface winds to 10,000 feet generally easterly in this area. Westerly winds usually signal bad weather.*

She glanced at the altimeter; it read 10,000 feet.

Might as well take advantage of the wind, she thought, as she pushed the control stick forward and brought the aircraft down to 6,000 feet. Around them stars disappeared as they dropped through a cloud deck. The aircraft shuddered as a gust of wind hit it. Amelia's arms tensed as she fought to keep the plane on course. Rain streaked the windscreen, illuminated by the cockpit's dull red glow. Outside was like a coal sack. She glued her eyes to the artificial horizon bobbing in front of her.

"Do you think this is wise?" Fred spoke matter-of-factly.

"We've got to make up some miles," she responded, a slight edge creeping into her voice as a strong gust buffeted the aircraft. "We've seen worse, Fred."

"Hang tight!" she said sharply, as the plane dropped suddenly in a vast air pocket. She poured on power, pulling back on the stick as the altimeter needle spun dizzyingly. Fred held his breath, hypnotized by the spinning dial. He sensed, rather than saw, the struggling woman beside him. As the needle slowed down, he let out his breath with a sigh.

"Some turbulence!"

She wasted no time answering him. Instead, she used precious fuel bringing the aircraft back to a safer altitude. At 6,000 feet she leveled off. "How much further, Fred?"

"Hundred fifty, two hundred miles. It's hard to tell with this wind and no stars to sight."

The cockpit lighted up brilliantly. A second flash illuminated a gigantic thunderhead towering in front of them.

"Better avoid that one," Fred advised.

The artificial horizon tilted right. Noonan's pencil rolled off his desk as the aircraft banked left. Again, the clouds lit up. From the picture frozen in his mind, Fred could see they were flying through a clear valley between two massive thunderheads.

Suddenly, the whole sky flashed around them; the plane jerked hard enough to clear Fred's desk. Swerving out of its left turn, the aircraft banked sharply to the right, almost standing on its wing. Inside, Amelia struggled desperately to regain control. She flicked her eyes across the gauges in front of her, already knowing what she would find.

"We've lost the right engine, Fred!" She glanced over her right shoulder, eyes big and round. "More than that," she added through clenched teeth. "I think the wing is damaged as well." She fought to keep the control stick from pulling forward out of her grip. "Strap in, Fred. This is going to be rough!"

As he struggled with his straps, another brilliant flash filled the sky. Once again, the whole plane shook, but when the flash was gone, the right wing continued to flicker.

"I have to ditch her!" Amelia's voice sounded shrill in the noise around them; it contained a hint of fear. Fred reached out and gripped her shoulder. She turned and saw his grin in the reflected glow of the instrument panel, highlighted by the flickering from outside.

"You're the best, Doll!" He winked. "Dinner's on me as soon as we hit Honolulu."

The damaged wing prevented her from leveling off. The best she could do was to keep the plane's spiral from becoming too steep. The altimeter blurred. They both began to hear the sound of driving rain against the aircraft skin; the flickering on their damaged right wing disappeared. Again, the sky flashed. Frozen before them, tumultuous waves stretched several hundred feet below. Amelia Earhart pulled back mightily on her control stick, trying to bring the nose up. She wrenched the wheel to her left, jammed her left foot forward.

Neither she nor Fred Noonan saw the waves rush up to meet them. Their damaged right wing caught the water first, flipping the plane up and cart wheeling it to the right.

"Fred!"

He lost his grip on her shoulder as the aircraft struck. Darkness closed in around them, but they could not feel the ocean pour in through the smashed windscreen. In the silence that followed, they felt nothing at all.

※

Sixty miles to the northeast a small group of men glanced from a lantern-illuminated runway on tiny Howland Island to a lightning-outlined squall off to the southwest. All they saw was a group of lightning-illuminated thunderheads...but as long as they lived they would remember.

Slingshot

PART I

"He shall have dominion over the fish in the sea…"

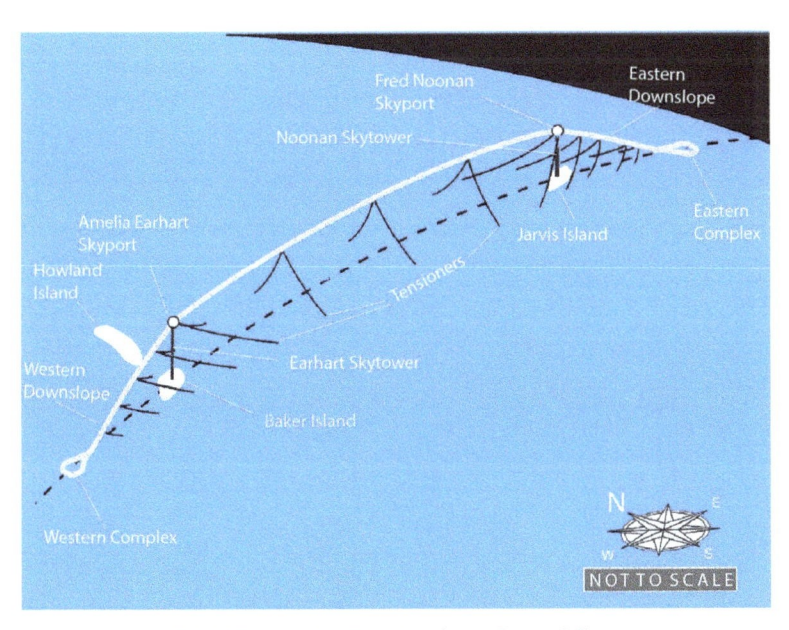

IMAGE 2—The Slingshot Space Launch Loop

CHAPTER ONE

EQUATORIAL PACIFIC—SOUTHEAST OF BAKER ISLAND

Margo stopped kicking her feet as the ominous gray shapes flashed into her peripheral view. Long, tawny hair floated past her head as her feet dropped below her slim, brightly clad body. She took a deep breath and floated slightly upward. A hint of fear crept into her mind as she turned toward three gray, sleek predators cruising just inside the limit of her vision, about twenty-five meters away.

A gentle touch on her shoulder startled her. She turned to see Alex Regent tapping the depth reading on his dive-console with his index finger. Margo reached down and grasped her console, turning it so she could read her depth: twenty-five meters. She had drifted upward five meters since seeing the sharks.

Margo exhaled angrily and let some air out of her breathing bag. She knew better than to lose track of her depth. Out there her life depended on a constant awareness of exactly how deep she was. Together she and Alex sank back to thirty meters. Off to their right, the three gray shapes drifted with them. Would she ever get used to it, she thought, as she released a bit of air into her bag to stop her descent.

"Alex," she said.

There was no response.

"Alex!" She tapped the back of her console several times.

"Alex!" Nothing but silence.

Alex placed himself in front of Margo and looked into her facemask. With his right hand he formed a circle with thumb and forefinger. His three other fingers extended straight up.

Margo returned the sign indicating she was all right while nodding vigorously. Then she pointed to her ear and lifted her console, tapping the back. Alex fumbled at his ear, and then tapped his console, and then shook his head.

Great, Margo thought, *EFCom is busted just when we really need it. Not busted,* she corrected herself, *just a submerged antenna.* She pointed to the three menacing shapes off to her right. Alex turned and scanned around them. Above and just behind them the blue-painted hull of their boat bobbed in the gentle waves. About twenty meters ahead of them hung a smooth, horizontal fluorescent orange tube about one meter in diameter. To the left it stretched into the gloom; to the right it angled downward. The fluorescent tube was attached to a slender cable angling up to the shadow of a buoy just beneath the surface to their right. Alex turned back toward Margo, making an exaggerated shrug.

Margo reached for her dive-console again and pressed a button located prominently on its face. The three sharks turned and commenced a meandering movement toward the two divers. Their front fins extended stiffly downward at about forty-five degrees. Their backs arched slightly, and their blunt snouts moved back and forth as they approached.

Margo felt her hair stand up on the nape of her neck. She turned to Alex and motioned him to her side. Alex withdrew a telescoped baton from its holder at his waist and extended it to its full one-and-a-half-meter length. He checked the safety lever near its handle, and with his thumb he flicked the lever so it pointed forward. As the sharks drew nearer, he held the stick out in front of him, pointed in their direction. Margo glanced around them again and pushed her console button once more. Alex waved the stick about slowly, and then steadied up on the nearest of the three menacing monsters.

Suddenly, with blurring speed, the nearest shark attacked. Alex struck out with his stick, the jolt of its impact rocking him backwards. A sharp crack was followed by a hissing sound as carbon dioxide rushed into the shark's body. In the same moment, flashes of silvery-black streaked from several directions. One of the remaining sharks was struck broadside by a dolphin's blunt nose. In a flash, it disappeared.

The animal Alex had injected rolled on its side and began a crazed, uncontrolled spiral toward the surface thirty meters above them. On its way up, it was hit several times by charging dolphins. It expired of massive embolisms before reaching fifteen meters. In the melee, the third shark vanished.

Margo reached out for Alex, grabbed a handful of breathing bag, and pulled him close to her. She placed the flat of her full-facemask against his and looked deeply into his eyes, as close to a kiss as she could come under the circumstances. Even down here they were deep blue. Several bubbles escaped from the positive pressure maintained inside their masks and shimmered their way toward the surface, expanding rapidly as they rose.

Like an old-time scuba diver, Margo thought, watching the rising silvery spheres. Instinctively she checked the volume in her breathing bag and glanced at the gauge on her tiny, ultra-high-pressure air flask. She found she was holding her breath, and as she felt the need to breathe, a gentle pressure developed against her back. She pulled back and turned to confront a two-and-a-half-meter-long dolphin nudging her from behind.

It was one of four that had responded to her sonic signal—George, her favorite. The other three dolphins crowded in around the neoprene and nylon suited divers, jostling each other for attention. Margo rubbed the head dome of each and indicated to Alex that he should do the same. Then the two of them turned their attention back to the tube suspended in front of them.

Alex swam to the angled portion and began to search along the tube's length, descending slowly. Margo dropped her arm from George's neck and kicked in Alex's direction, keeping him in sight, but staying between him and the surface. The four cetaceans arrowed toward the surface and grabbed a gulp of air, then settled back down, playfully cycling between Alex and Margo, gently jostling them. About thirty minutes later, Alex motioned Margo to join him. She released a bubble of air from her bag and dropped down beside him. Her console showed a depth of fifty meters. Alex pointed to a five-centimeter rip in the bottom curve of the tube's fluorescent covering.

Margo reached into a deep pocket located on the left leg of her suit and withdrew a role of patching tape. Alex stretched the edges of the tear, and Margo applied a strip of self-sealing tape along the opening. Then she located a small pneumatic valve on the top of the tube and attached a hose from her spare air tank. On a signal from Alex she released air into the tube, forcing water out through a one-way valve on the underside. She stopped when bubbles escaped from the lower valve.

As the tube rose slowly, Margo held on, keeping track of their progress on her console. They stopped rising when the gauge read thirty meters. Margo felt the tube—it was taut and solid. She tapped the back of her console, listening for the faint rush of sound in her ears. Nothing. She pointed to the back of her console and then her ear, and shook her head. Alex offered another of his exaggerated underwater shrugs and grinned, although the only part of the grin she could see was his crinkled eyes. She grinned back and pointed toward the suspension buoy and their boat, making an angled upward sign with her free hand. Alex nodded, checked his console, and they both headed back, slowly rising as they swam.

Margo saw Alex check his console from time to time, making certain they kept below the ever-changing ceiling limit it calculated for him. Since she had remained shallower than Alex for most of the dive, she knew she would be safe following his lead. She looked around at the four dolphins. Her earlier fright was gone, and she simply enjoyed George's protective nearness and the playful bumps and nudges from the others.

On the surface finally, Alex dropped his facemask down around his neck, fully inflated his bag and grinned at Margo. "Close call down there!"

Margo shoved her facemask down and patted the glistening snout that appeared in front of her. "Thanks, George. I love you too."

The dolphin mewed a pleased response, lifted his body out of the water and backed away, chattering as he went. The other three animals circled at and below the surface, keeping watch over their human charges.

"What happened to the EFCom?" Margo asked. "I expected it to come back on line as soon as the antenna surfaced."

"Broken antenna wire, I imagine," Alex answered.

"Storm damage, I'm sure," said Margo, as they turned and headed toward the waiting vessel.

"Probably," agreed Alex. "But that wasn't a burst seam," he added.

"Yeah, maybe the sinking tube snapped the wire."

Actually, tube flotation chambers flooded on a regular basis. They had patched a full ten percent of them since the project started. But it was a bit unusual to find a rip on the tube bottom, and the Electrostatic Field Communication ("EFCom") transceivers on the buoys almost always survived.

※

The EFCom buoy nearest the tear had ceased transmitting, and the buoys on either side of the tear had signaled their departure from datum a day earlier. Alex had opted to employ an electrostatic field communication system, because of its clear underwater signal transmission capability that was independent of acoustic conditions, since it didn't rely on sound transmission through the water. Every buoy, each skimmer and floater, and every diver was outfitted with one of the small EFCom transceivers. Alex had inspected the non-transmitting buoy personally during an overflight from Jarvis Island. There was nothing visible on the two kilometers of surface between the buoys; they were closer together, but not so that it was visible to the eye. Nevertheless, the remaining 1,828-odd buoy-suspended kilometers of tube were stressing from the downward pull of the waterlogged section. The buoy near the tear was several meters underwater.

Suspended inside the flotation tube were two virtually impervious, lightweight, hose-like tubes, each about six centimeters in diameter, called vacuum sheaths. Two shallow channels jutted out from the bottom of each vacuum sheath, filled with electronically controlled suspending magnets. Magnetically suspended inside each vacuum sheath was a five-centimeter tube of segmented soft iron officially called the rotor, but more popularly known as the ribbon, so named from the earliest conceptions back in the 1980s of the Launch

Loop inventor, Keith Lofstrom. Alex was eager to check continuity readings to make certain the vacuum sheaths had not breached. They were not yet evacuated, but seawater entry at this stage would seriously delay the entire project. If the EFCom had not crapped out, the tests would already be underway.

Alex glanced ahead at Margo Jackson, cavorting with her four dolphins as they made their leisurely way back to the waiting boat. His field engineer in charge of underwater construction was a remarkable female. Nearly as tall as his own 183 centimeters, her model's slender figure, encased in electric-blue nylon-covered neoprene, seemed to lack feminine curves. He knew differently, of course, having joined her bikini-clad person from time to time for morning swims since the project began over two years ago.

The project—Alex had lived with it for three years before actual construction began. Longer, actually, if you considered dreams—since before the incredible, worldwide bi-millennial celebration, when he still was a young boy.

There was the nearly simultaneous publication in America and England of practically identical ideas in 1985. Paul Birch published an article in *The Journal of the British Interplanetary Society*, while in America Keith Lofstrom published his article in a supplement to *The Journal of the Astronautical Sciences*, he recalled. Nobody could agree on the names: Skyrail, Launch Loop, Beanstalk. There were others, but the idea is what counted, the sky-shaking idea that you don't need rockets to get into space.

Newspapers were full of explanations three-and-a-half years ago when the aging president of a computer software giant made the announcement. He would funnel a significant portion of company profits into the consortium. Space travel would become as commonplace and inexpensive as the personal computers his pioneering work had made possible. He went on to outline the easy-to-understand concept.

Imagine a water hose streaming water in a parabolic arch. Deflect the water and funnel it back to the start through a pump, creating a closed system. Make the stream strong enough and the hose light enough, and the entire structure will support itself—the water holding up the hose structure. Now, replace the water with a thin, closed-loop pipe of segmented soft iron. Make it 5,000 kilometers around and accelerate it to orbital velocity with gigantic linear induction motors from two points on the equator 2,000 kilometers apart. The center section of the structure, including both the outgoing and return legs of the loop, will rise to about eighty kilometers above the Earth. Supply access to the upstream end in space with a Kevlar-hung elevator, and you can launch capsules by magnetically coupling them to the rapidly moving pipe of iron.

Slingshot, they called it. The greatest engineering undertaking in the history of the world, they said.

As the on-scene project manager, Alex was responsible for getting the job done, on schedule, on budget. He was building a gossamer structure over 2,500 kilometers long, a frail spider web, completely invisible when viewed from more than a few kilometers. Alex grinned wryly. All *Slingshot* really consisted of was a fancy evacuated tube, a flexible iron pipe, four linear drivers and their power sources, some guy wires, and a couple of elevators. Put that way it seemed simple enough. But, of course, it wasn't simple at all, and for all his skill and engineering competence, and despite surface appearances, deep down Alex was not entirely sure that he could make it happen.

Margo and Alex climbed up the ladder and onto *Skimmer One's* bobbing fantail. This was one of two skimmers on the project—twelve-meter-long surface-effect boats that looked more like a floating aircraft than a traditional motorboat. They were capable of 200 knots, skimming about one-and-a-half meters over the wave tops. They had a small open fantail, just large enough for a couple of divers to doff their gear. Being on the fantail when the skimmer was on its cushion was more than dangerous, and was strictly prohibited throughout the project.

Alex signaled to the waiting coxswain and they got underway for Baker, plowing through the water while Alex and Margo remained exposed. He and Margo stood near the stern railing and removed their dripping skins. Alex looked back at the buoys, now presumably in their proper places.

"How many more times?" Alex looked quizzically at Margo.

"Who knows?" She glanced back at the bobbing buoys. "We have repair people available at both ends. We shouldn't be doing this ourselves, you know." She turned and looked directly at Alex. "What do you think—weather or sabotage?"

Alex shrugged and tossed the spent carbon dioxide cartridge from his shark stick in the general direction of the cavorting dolphins. "I wanted to see for myself, and I still don't know. Does it matter? We can't patrol the entire eighteen-hundred-twenty-eight kilometer length anyway."

"What are we dealing with?" Margo asked. "You don't get out here in a rowboat."

"We're two thousand wet klicks from any kind of civilization," Alex said. "At minimum, that's a large motor-yacht or even an ocean sailer—you know, one of those we maybe can afford when this job is done." He sighed. "We're dealing with lots of money and someone with a major bitch."

He looked into her green eyes.

"Just keep my tubes at depth." His blue eyes flashed, and he turned toward the cockpit to radio his orders to test pipe continuity.

※

Margo dropped her eyes at his challenge. For the thousandth time she asked herself if she had bitten off more than she could chew with this assignment. Was it her fault that the flotation chambers kept ripping? Was she missing something important? Was she copping out to imply there might have been sabotage? And yet, Alex seemed to agree that it might be sabotage. When she joined the project two years ago, the newspapers had acclaimed her as the ideal role model for the new twenty-first-century woman. At times that burden lay heavily on her shoulders, as it did now, she reflected.

It was a vast responsibility, and there was no way one person actually could control all of it at once. How Alex handled the weight of the entire project awed her, but she was careful never to let him know.

Margo watched Alex step into the cockpit. He was tall and slender, richly tanned from his constant outdoor work. She felt a softness well up inside her, a gentle warmth spreading out from the pit of her stomach. She bit her lower lip and turned angrily to lean on the after-railing.

None of that, she chided herself. This assignment was too important, and the stakes too high, to let any kind of emotion intrude. As she entered the cabin and sealed the port, the skipper switched modes, and pressurized air quickly filled the hard-sided skirt. In moments the skimmer lifted out of the water, except for the port and starboard skirts that protruded about a meter into the waves. Within seconds, high-pressure water nozzles jetted water from the end of each skirt, and within thirty seconds *Skimmer One* was approaching 200 knots.

As *Skimmer One* headed into the afternoon sun, trailing an arrow-straight wake of white foam, Margo stood looking aft through the sealed port, remembering her instinctive sharing, and their underwater kiss following the fright of nearly becoming shark food. She shook off the sensation and busied herself with putting away their diving equipment. But a hint of a smile remained on her lips as they shot over the surface, finally settling back onto the water as they entered the small protected artificial harbor on the west side of Baker Island, just south of a shallow reef that went dry at low tide.

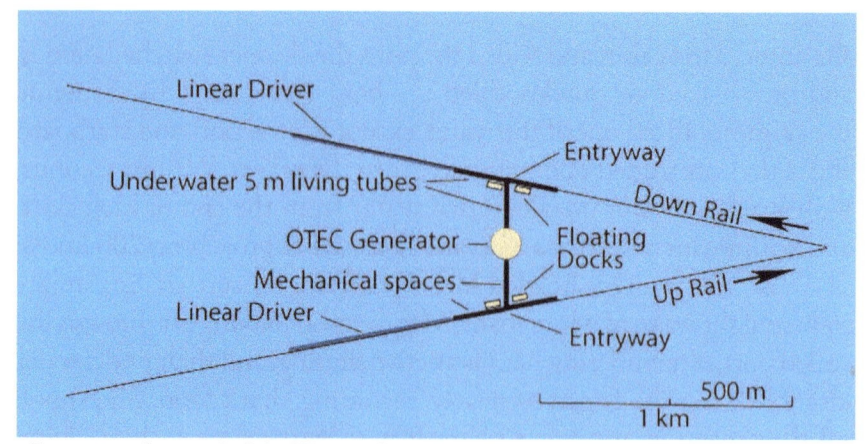

Image 3—Apex end of the Western Complex—The two arms of the Complex meet at the far side of a twenty-kilometer-wide circle about fifty kilometers further west

CHAPTER TWO

WESTERN COMPLEX—300 KM WEST OF BAKER ISLAND

"Hey, Alex! Got control of those tubes yet?" Klaus Blumenfeld stretched out a massive helping hand, his gray eyes twinkling.

Alex glanced at the big man's uncovered, shaven head. "You're going to burn that top yet, Old Boy."

He stepped down from the floater's cockpit to the pontoon and across to the dock, matching the pace of the tall German engineer as they made their way along the floating catwalk that marked the underwater location of the tube complex, toward a protruding hatch at the intersecting walkway in front of them.

"Like the *Schimmelreiter*," Klaus said with a grin.

At Alex's puzzled look he added, "You know, the ghost rider in Thomas Mann's famous novel."

The hatch did resemble a hooded ghost's head, looming up from the catwalk, its recessed door lost in dark shadow.

The wheel on the massive oval submarine hatch turned smoothly. Klaus swung the door out and motioned Alex to precede him down the steep staircase or "ladder," as Klaus insisted on using nautical terminology. Once inside, he closed the outer hatch and joined Alex at the bottom of the ladder some five meters down. He glanced at a monitor panel on the bulkhead, and when the lights had all turned green, opened the hatch in the wall. Alex and Klaus passed through the hatch into a small chamber with a large oval pressure-door, that Klaus called a "seal," set into the opposite wall. Klaus palmed a plate on the bulkhead and, with a faint hiss of compressed air, the seal twisted slightly and slid back into the tube wall.

They stepped through the opening into a chamber that served as both a pressure lock and elevator. To the right were three plates: S for Surface, L for Lock, and T for Tube. The S glowed with a bluish light; Klaus touched L, and the seal closed behind them. A rush of air greeted them. Alex swallowed to clear his ears, and Klaus just yawned and touched T. Alex felt a slight downward motion, and in about thirty seconds the seal hissed open to reveal the spacious living quarters suspended halfway between the surface and the massive two-kilometer-long linear induction motor Klaus was constructing.

They stepped out into a thickly carpeted passageway, a five-meter-diameter tube that stretched directly out from the combination pressure lock and elevator. Another five-meter carpeted tube stretched to the left. Every few meters large portholes pierced the curving bulkheads, so that the passageways were bathed in eerie, flickering green light. They strode down the left tube, past several portholes and turned right through an open pressure seal into a small chamber. Klaus palmed a plate on the bulkhead. With a faint hiss, the seal behind them closed, and a few moments later one in front of them opened.

"Safety first!" Klaus grinned at Alex as they stepped through the emergency air lock into his personal living quarters. Alex had ordered that all living quarter air locks be interlocked to close automatically should one of the interconnecting tubes breach.

The wall before them was entirely transparent. It looked out over the construction activity through water so clear that it hardly seemed to be there at all. Off to the right, divers assisted a submersible in emplacing a section of tubing. Their back-mounted twin gills twinkled with reflected morning sunlight. The main tube stretched to the right, vanishing in the watery distance. It was punctuated by periodic blisters that formed individual living modules currently occupied by the construction crew, but slated for power plant personnel once the system became operational. To the left, just visible through the transparent wall, their tube extended for 200 meters to intersect the massive vertical Ocean Thermal Energy Converter, or OTEC cylinder as they called it, that started four meters above the surface and ended a thousand meters below. It was a hundred meters in diameter and could be seen clearly through the transparent water to the left.

Alex stood gazing through the wall at the frenzied panorama before him. A classic Beatles song floated through the slightly compressed air surrounding them while Klaus built a head on the beer he handed to Alex.

"Easy to make a head at this depth," he said with a grin, wiping the foam off his upper lip with a massive hand. "But you should try it at one hundred meters! The bottle sucks air in when you open it!"

"Come on, Klaus...."

"No, really. At that depth, room pressure exceeds bottle pressure. It sucks, let me tell you!"

Both men laughed and Klaus sat at his desk, turning his chair to look out the transparent wall, while Alex sat in an overstuffed chair facing Klaus.

"Well," said Alex, "Margo got the tubes back to thirty meters this morning. Lost anchor buoys are giving us more problems than I anticipated, you know." He crossed his legs. "I'm beginning to come around to Margo's perspective that there's more than ocean waves and coincidence at work here."

"Are we talking pirates and peg-legs?"

"How about ocean sailers with big money backing?" Alex sighed, feeling the weight of his job. "What's your situation now?"

Klaus reached behind the desk and pulled out a roll of blueprint drawings. He could have called them up on his Link, but he preferred the substantial feel of the older drawings. Alex never could understand the real reason for Klaus' anachronism, but Klaus was the best, and the paper came with the territory.

Klaus unrolled the drawings on the deck between them and anchored the curling edges with their beer mugs. Several dark circles on the drawings testified to Klaus' usual mode of work.

"Here we are," said Klaus, his large hand spread out over a tear-drop-shaped circular feature with its large end about twenty kilometers across and its smaller end drawn to a point about sixty kilometers to the east. "We've been lucky, you know."

Alex raised an eyebrow.

"Look at it this way. My work area at this end is spread out over more than six-hundred square-kilometers...."

"Wait a minute!" interjected Alex. "Ninety-five percent of what you do takes place right here." He pointed to the small area in the apex of the point. "That's one square kilometer at the most...give me a break!" He picked up his beer and took a swig. "And most of that is right out there." He gestured at the activity taking place outside the transparent wall.

"Okay, Alex, okay. But let's assume a man is hurt right here." Klaus pointed to the farthest section of the semicircular deflector. "At best, that's a half hour trip, even using the skimmer." Klaus jammed his finger first at the westernmost end of the deflector and then at the downslope at the structure apex where the two linear induction motors were located. "I need two floaters, Alex, two of them!"

"Klaus, Klaus...the cost! Two planes, the pilots, fuel—that's a small fortune, man. Don't you know that?"

"*Ja*! And make one of them a *Fräulein*!" Klaus grinned and bumped mugs with Alex. "To the *Fräulein* pilot!"

Klaus let the top drawing roll up. The second one was a detail containing just the semicircular deflector, a thirty-two-kilometer-long rigid structure formed of powerful electromagnets designed to deflect the rapidly moving tube of segmented soft iron in a 180-degree reversal.

"The electromagnets are going in place this week; actually, should be done by day-after-tomorrow." Klaus got up and walked over in front of the transparent wall, his back to Alex. "We can test continuity then, but we're still behind schedule on the well. Give me more men," he turned and grinned at Alex, "and I'll do it faster."

Alex exposed the next drawing. It showed a plan view of a hundred-meter-wide Ocean Thermal Energy Converter with its attendant surface structures. It also showed a foreshortened orthogonal view of the entire structure. Alex worked with it every day, but it still boggled his mind. The massive ferro-cement tube extended a kilometer down into the deep, a gigantic well suspended in the ocean. Surface waters in the well heated by the sun's rays expanded convectively and flowed out the structure through spillways at the surface. This water was replaced by dense cold water that flowed up through the tube from one kilometer below, turning large turbine blades as it passed, generating massive amounts of power. Power to drive the linear induction motors at this end of *Slingshot*, to bend the ribbon in its path around the semicircular deflector. Sufficient power to supply a medium-size city, almost five-hundred megawatts, like a nuke plant or a large coal plant.

"Have you experienced further problems with the well?" Alex pointed to the orthogonal view of the massive generator.

"Not since the bad shipment of syntactic foam, the batch that collapsed under pressure last week." Klaus turned around and shrugged. "What the hell, Alex! Nobody's ever done it before. Bound to be some problems, what?" He grinned. "Now, what about those extra men? I really need them to get back on schedule."

"I can't, Klaus. I'm sorry, but there's just no way."

"Not even using the money we saved with my modifications?" Klaus' shoulders drooped a bit.

"Klaus, what you did was magnificent. No! I really mean that. It really was!"

Klaus tried to interrupt.

"Hold on, friend. You conceived of using syntactic foam spherules as binding material in the ferro-cement. That was a stroke of genius. It's made the difference between staying on schedule and budget instead of whatever-the-hell." Alex smiled warmly at his friend. "But there still is no way I can hire on more underwater workers." Alex held up his hand as Klaus was about to speak. "I may be able to find two or three in Margo's crew…"

Klaus burst into smiles, the sag gone.

"On loan, of course, against your budget."

The German's smile remained where it was. He held up his nearly empty mug. "*Danke*, Margo. I love you!"

Outside, the submersible backed away from the emplaced tube section. The divers disappeared through a hatch in the tube's side. The faint sound of rushing air could be heard inside the snug living room where the men continued their discussion. Unseen from their vantage point, water rushed out through wells extending through the tube's underside. When bubbles started pouring through these openings, the rushing stopped. A school of brightly colored fish chased the flashing bubbles swirling up the curved tube walls, but they stopped short as the silver spheres broke free and headed toward the surface.

"I'll have to inspect that tube section now," said Klaus. "Those boys are good, but when they know I'm looking over their shoulders, they're even better." He grinned apologetically. "You can come, if you want."

Alex knew better than to take him up on his offer. This was the German's area of responsibility. His men had to know that the buck stopped with Klaus.

Alex stood to go, leaving the blueprints on the table. "I know my way out." He headed for the air lock.

Klaus turned toward the opposite wall, looking over his shoulder. "I'm going to swim over. Keeps them on their toes, not knowing where I'll show up next."

He passed through a doorway as Alex stepped into the lock. As the seal slid into place, Alex saw Klaus laying out his skins and gills.

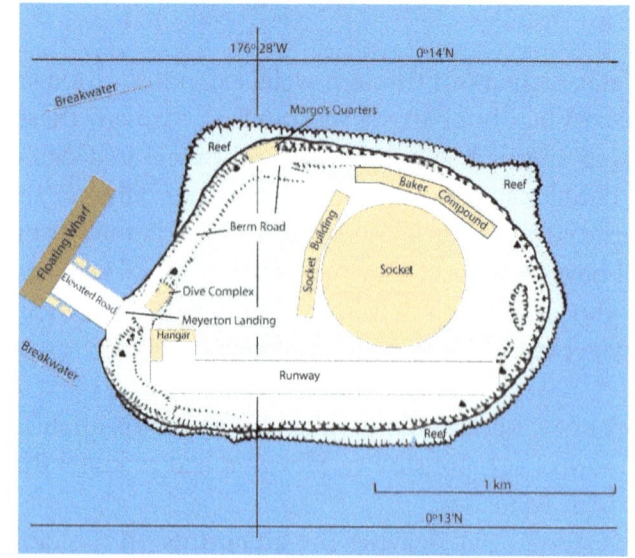

Image 4—Baker Island showing locations of the various facilities

CHAPTER THREE

BAKER ISLAND—MARGO JACKSON'S QUARTERS

Margo stood in front of a large picture window surveying the panoramic spectacle of long Pacific rollers breaking against the exposed coral reef on the north-west side of Baker Island. Her quarters were housed in a low, chalk-white building perched atop a bluff five meters above a rugged tidal flat. At high tide, breakers rolled across the shallow reef, churning white here and there, but always reforming, picking up momentum, to dash against the sun-bleached coral barrier below her window. But now, across the rugged expanse of brown coral, living polyps struggled to survive their exposure to the noontime tropical sun while the long breakers spent themselves against the reef boundary 200 meters further out.

Margo's gaze focused on the horizon. Sixty-seven kilometers beyond the breakers lay Howland, and some three thousand kilometers beyond Howland lay Honolulu. *Amelia's next leg*, she thought, and turned to the framed photo on the wall behind her.

"What happened that you missed Howland? Where are you now?"

Margo was troubled by the puzzling disappearance so many years ago. *TIGHAR* claimed to have solved the mystery back in the nineties. Members of *The International Group for Historic Aircraft Recovery* had discovered artifacts on Nikumaroro Island that eventually convinced them Amelia Earhart and her navigator, Fred Noonan, had somehow reached its shores, some 550 kilometers south of Howland. Margo was not overwhelmed by the evidence. Noonan was one of the finest aerial navigators in the world. His last message to the U.S. Coast Guard Cutter *Itasca*, standing off the eastern end of Howland, was: "KHAQQ calling *Itasca*. We must be on you, but cannot see you…gas is low…."

To Margo this did not sound like they were hundreds of kilometers off vector. She watched a line of squalls passing the island several kilometers out. Thunderheads dominated the pattern, moving with the squalls like sentinels of doom. A bolt of lightning flickered between two looming cloud towers. Then a flash caught her eye, and she squinted against the noontime glare to make out a small plane banking toward the island.

"Be careful, Alex," she whispered as she turned to her Link.

With a sigh, she slipped into the straight-backed chair in front of her Victorian-style desk and tapped her Link pad. The space above her keyboard shimmered momentarily. Sparkling points of light coalesced, and then the wall behind her desk faded behind a three-dimensional image of the 1,828-kilometer-long submerged tube complex that had become her sole reason for existence.

The orthogonal birds-eye view arrived at her Link courtesy of GS32, hanging in orbit some 40,000 kilometers overhead. The occasional sparkle on the translucent ocean surface was actual sunlight glinting off a wave. Eighteen-hundred-twenty-eight buoys appeared as tiny specks identifiable only because they formed a continuous line stretching between Baker and Jarvis Islands.

Margo touched her pad and a faint grid appeared on the ocean surface. The Equator, glowing red, crossed the line of buoys at a shallow angle to the east about halfway to Jarvis Island. She touched the pad again, and the picture zoomed out, revealing the east coast of New Guinea, and the Port of Lae, some 4,100 kilometers west and slightly south of Howland. And Nikumaroro Island, about 600 kilometers almost due south. She moved her cursor to hover near Nikumaroro.

"Did you end up here, Amelia?" she murmured. "Is this where you spent your last days, while the world searched up here?" She moved the cursor north to the wide end of a purple wedge that started at Lae and terminated with a 500-kilometer-wide cross section centered on Howland. The wedge contained Baker Island and the growing deflector complex west of Baker. Margo zoomed in on the complex.

Her Link-enhanced satellite image clearly showed the large circular opening of the thermal generator. She tapped her pad, and the Link superimposed a shimmering structure around and above the generator. It looked like a symmetrical teardrop with its point located 330 kilometers west of Baker. The OTEC generator was nestled in its point. A faint line exited the point, angling into the sky at about fifteen degrees. A faint vertical line connected the middle of Baker Island with the sloping line eighty kilometers above the surface.

Margo tapped her pad to zoom the picture out. The faint structure followed the line of buoys for 1,828 kilometers to Jarvis Island, where it joined another vertical line emanating from that island's center, and then sloped down to a mirror image of the structure west of Baker.

It seemed so small and delicate when viewed this way. *Twenty-five-hundred klicks; so vulnerable…to wind, to wave….* She paused in her thoughts and glanced out at the receding squalls. *…and to sabotage….* A worried frown settled across her face.

She had found the break on the underside of the tube. How did it get there? They were nowhere near normal shipping lanes, and besides, the break was on the *underside*; so how did it happen? For now she decided to write it off to natural causes, but she resolved to undertake a statistical analysis of all the breaks thus far. If it really was sabotage, she needed to know. But it seemed so unlikely. After all, who would care? Who could possibly benefit from damaging *Slingshot*? Even the Green community seemed to be supporting their project, perhaps because *Slingshot* would significantly reduce rocket propellant exhaust in the atmosphere. Sabotage seemed so unlikely; but the analysis would show something, she decided, as she turned to listen to footsteps crunching through the crushed coral walkway leading to her door.

The door opened, framing Alex's silhouette against the noonday brightness.

"Margo!" Alex grinned at her as he strode into the room.

A faint whiff of ozone mixed with aviation fuel accompanied him. He turned and looked out the picture window.

"You definitely have the best view for two thousand klicks in any direction." He spread his arms as if to encompass the receding thunderheads.

"Drink?" Margo asked, walking toward an antique cabinet against the wall.

"Got any more of that *Islay* you picked up last month in Sydney?"

Alex glanced at Margo's Link. "Earhart again…" He grinned at her as she handed him a snifter containing a clear dark amber liquid.

Margo shrugged. "The mystery isn't getting any less mysterious." She pointed to Nikumaroro Island. "How could they possibly have gotten down here, Alex? That's almost eight-and-a-half degrees off course." She was referring to the difference in vectors from Lae to Howland Island and Nikumaroro Island. "You certainly wouldn't make that kind of mistake, and Noonan was as good as they got back then."

Alex held up his snifter in a toast. "Maybe that's the key word… back *then*."

Margo finished pouring herself a glass of Columbia Gorge Merlot, and returned the salute; sunlight sparkled through the deep red liquid. She reached out and tapped her pad as Alex inhaled the pungent fumes collected in his snifter. The satellite view of *Slingshot* disappeared, replaced by a translucent, double-layered chessboard with a game several moves in progress. "Do you have time?" she asked.

"Depends," he answered, "on whether you or your Link made the last move." His eyes twinkled.

Margo feigned hurt and retorted, "How else can a girl challenge the MIT Hyperchess champion?"

She sat down in front of the mid-air image and tapped her pad. Most of her white pieces still occupied the upper field where they started when the game commenced. The white-queen-bishop-pawn faded from the white square it occupied on the upper field to the white square directly beneath it, where it faced an array of black opponents.

Alex studied the display for a while and then looked at Margo in admiration. "For a duffer, that's pretty sharp." And then he added, "That's not fair of me, Margo, that's a sharp move, plain and simple. I'm impressed."

He turned back to the display. After a short while of study he tapped his pad. The black-king-knight moved from its original black square to white-king-bishop-three and faded to the white square on the field above it. Alex leaned back and smiled, inhaling the fumes from his snifter before letting a few drops of the precious amber liquid engulf his taste buds.

Margo sipped her wine and leaned forward, frowning. She reminded herself that knights always faded—each move placed them on the opposite field *and* the opposite square color. All the other pieces either transited by fading to the opposite field, making a move, and fading back to the original field, or they just faded, but not both. She faded her queen-bishop from its normally black square to the black one below—so bishops retained their square-color dependence, but knights took on a color dependence determined by which field they occupied.

Alex's next moved placed his knight under Margo's king; a translucent cylinder appeared around her king. It was caged, and unless she could capture the caging knight with one of her own, she would inevitably lose the game. She studied the display for several minutes, and then she saw it. Instead of trying to capture the caging knight, if she could distract Alex for one move, she could cage *his* king in a mating move. But, of course, his game was so good that he probably would see the distraction for what it was. So…how about a second order distraction?

"Do you think we have a lot in common?" she asked him, holding her glass of Merlot up to the light, turning it, letting the sunlight illuminate its ruby redness. "I mean, Merlot is robust and hearty…you know, the connoisseurs turn up their noses; and *Islay*, well, *Highland* it's not." She tapped her pad idly, jumping her knight to the upper field, while her eyes twinkled as she offered him a toast.

"To taste…not *good* taste!" She touched her goblet to his snifter. They both sipped.

"Dirty pool, Margo!" Alex said. And he faded his king exactly as she hoped he would.

"Mate in three moves!" Margo said with an impish grin.

And it was.

✺

"Down to business."

Alex stood up and began pacing across the room, silhouetted against the panoramic view of the rising tide over the reef outside the window. Waves were beginning to break over the rough brown surface. Here and there boiling green water churned up through and across the exposed coral, only to be sucked down again between delicate branches of living rock. A faint mist formed over the entire shelf, as it seemed to sink ever so slowly into the sea.

"Klaus is close to activating OTEC West. He's having problems, too, you know."

Alex folded his hands across his chest and then stretched them over his head, palms extended upward. Margo's heart skipped a beat, and she forced herself to concentrate on business.

Alex continued, "Nothing definite yet; just a vague feeling that things are happening that shouldn't."

Margo nodded. "Tell me about it."

She got up and moved her lanky frame to the couch where she sat, propping her legs up. "I'm essentially ready for power, you know."

Alex looked at her and nodded.

"The rail's in place," she continued, "and properly suspended along the entire eighteen hundred twenty-eight klick path. The west deflector section is in place, and my people are working hard on the east deflector section as we speak." She swung her feet to the floor. "We're setting the west T-anchors now, and as soon as Klaus can get me power, we can start connecting the tensioners. Half of them are on Baker, and Klaus is holding the other half for me."

Alex sat down at Margo's Link and tapped the pad. The Hyper-chess game was replaced with a scheduling flow chart. He entered several notations, and then solicited more detailed information about the exact status of Margo's side of the project. Margo pulled up a chair near him, and for an hour or so they concentrated on the details of the underwater elements of *Slingshot* for which Margo was responsible. She tried hard not to respond to Alex's closeness.

Suddenly the display flashed, and a three-dimensional image of Klaus' polished pate appeared in the upper-right corner. Margo pushed her chair back and stood up, moving out of the Link scanner's range. The move was instinctive; she didn't know why she did it.

"Alex…I was expecting Margo…Is she there? Are you guys alone?" the simulacrum asked.

"Margo's here. What's up?" Alex waved Margo to the display.

"I need you out here, Alex. Something has come up…and bring Margo. I think she needs to see this."

"What is it?" Alex's voice sounded clipped.

"Not over the Link, Alex."

The entire operation used normal security precautions. Link connections were routinely encoded to level three, more for data integrity than anything else, since the correlation circuits at level three virtually ensured that no transmission could be lost. Although it was widely believed that security at this level was complete, insiders knew that a determined expert could hack into the system, but could not change anything without leaving evidence of the intrusion. It was odd, though, that Klaus was unwilling to discuss his information over the Link.

Margo's thoughts flashed to the torn tube.

"Hi, Margo!"

The simulacrum smiled in Margo's direction. It was a remarkable illusion, she thought, as she answered back.

"Hi, yourself, Klaus!" She smiled self-consciously. "We're on our way." She lifted her eyebrows at Alex.

He nodded.

"Have a micro-brew ready," she ordered good-naturedly.

BAKER ISLAND—AIRSTRIP

Alex popped an *Alim* on his way out the door. "No sense taking chances," he said, as they hopped into a white pickup and drove along the berm to the airstrip still left over from before World War II, but enlarged and fully modernized now.

By the time his engines started up, any residual alcohol from the scotch he had consumed had been converted to simple sugars in his liver. He taxied to the end of the strip on wheels extending through the plane's pontoon floats and revved his engines.

CHAPTER FOUR

AIRBORNE TO WESTERN COMPLEX

Normally Alex would have taken the skimmer, but at its best it could do about 200 knots, making the Western Complex about an hour away. Since his amphib cruised nearly twice as fast, there was no contest. Klaus' urgency dictated the fifty percent savings in time. Alex banked right out of a steep climb and headed just south of west. He pressed a button on his console, and a flashing pip appeared in his heads-up display just left of the centerline indicator. He adjusted his course slightly and set the autopilot.

The sky was pale blue and shimmered somewhat in the noontime tropical sun. The passing squall had left a fine mist in the air that scattered the sunlight, softening the line where sky met sea, surrounding them with an ethereal glow.

※

For Margo the moment took on a special significance. The world below seemed distant, the problems remote. Her immediate present was out of her control, but pleasantly so, in the hands of this complex man she sometimes thought she knew well, and other times discovered she hardly knew at all. She glanced over at Alex as he set the aircraft on its course and settled into a relaxed state of attention. She knew well her own problems and responsibilities for completing the underwater elements of *Slingshot*. But when she considered the load that Alex carried, and as she watched him in action, she was filled with boundless admiration. He was a relaxed man who made you feel comfortable in his presence. Responsibility for building the entire multi-billion-dollar project rested squarely on his shoulders. It was an awesome load, and yet he retained a boyish good humor and a sense of adventure that completely belied the scope of his responsibilities. In a word, he instilled confidence—in himself, and in the people under him. Margo was, frankly, awestruck by the man. But Margo was unabashedly good at what she did, which is why she was selected over a large cadre of highly-qualified applicants. Nevertheless, she harbored a niggling doubt that the fact she was a woman probably influenced the decision. In her mind's eye she could follow the boardroom level discussion back in Seattle headquarters, occupying the top four floors of the historic Smith Tower.

"Hey John, get a load of this applicant." He slides the portfolio across the polished tabletop, glancing surreptitiously at the lone female board member sitting at the end of the table. It's obvious to the other male board members that he is not saying what is really on his mind. "She's as good as they come, and she's female." He looks directly at their woman colleague. "What do you think, Mabel?" as John slides the portfolio down the table to her.

Mabel had fought her way to the top, tooth and nail for twenty-five years. There was nothing these fellows could put over her. "Settle down, Rex. This is business," she snaps as she examines the accomplishments of the tall, disgustingly good-looking, incredibly accomplished young female engineer. "Who are your other finalists?" She directs the question to Rex, who slides three other files across to her, all of them male applicants.

"Any one of these four can handle the job," John comments.

"The power team and the overall project will be run by men," Rex added. "Everything else being equal…"

"That's crap, Rex, and you know it!" Mabel gave no quarter, which was probably a significant element in her rise to the top. "Which of these is the best?"

"The differences are minor, but," he selects two files, "this and this…" The female applicant was one of the two.

"Okay," Mabel responded, laying the open files flat in front of her, "now, point by point…"

Margo closed her eyes and winced, remembering Rex Johnson's comment when she appeared before the board to accept her selection.

"Really glad the best choice was a woman," he had said. And Chairman Mabel Fitzwinters had smiled and passed her that knowing look that women exchange when their male colleagues make asses of themselves.

I really don't know, Margo thought as she transfixed her gaze on the cocooning, shimmering mist. *I really don't know if I'm here because I'm the best, or if some poor schmuck lost out because he had a cock between his legs.*

"Margo?"

Alex turned toward the willowy blond next to him. "Hey, Jackson…you all right?"

Margo blinked her eyes and, with a small start, gave her head a little shake.

"Sure…just thinking…" Her voice trailed off. "What an odd day," she remarked, looking out into the translucent mist. "Good thing it's not like this every day."

"Yeah, without this," he indicated the heads-up vector display, "a fellow could get lost."

"They were flying at night," Margo retorted, "and besides, they did have a radio vector on the *Itasca*." *But they left their new radionav gear in Lae*, she added to herself as she looked out the aircraft window. *But eight and a half degrees…* She just couldn't buy it. *Noonan had too many air miles under his belt. You didn't live long making mistakes like that; but, of course, they had not made it.*

WESTERN COMPLEX—300 KM WEST OF BAKER ISLAND

Klaus watched Alex bank the sleek craft into a tight turn and drop low over the apex of the western deflector. From above, he knew, the water was so clear that it was difficult to distinguish which parts of the complex were submerged. The huge OTEC cylinder wall along with the tubes leading to the apex formed a virtual harbor. The tubes offered some protection from the prevailing winds that blew from the east, so the water in the harbor was relatively calm. Alex landed and taxied to a floating dock where Klaus was waiting with a friendly grin and a wave.

Alex cut the remaining engine as the aircraft drifted alongside the dock. With practiced expertise Klaus whipped a line around the forward cleat and figure-eighted the bitter end around the cleat at his feet. He repeated the task at the aft end of the pontoon while Alex and Margo deplaned. No one else occupied the dock in keeping with one of Alex's fundamental on-the-job principles: *Your load plus ten percent.*

Every individual hired by *Slingshot* was highly compensated, and each position from Alex right down to the power-room grunts was filled by the best available talent. There were no perks, in the traditional sense of the word that set off one position from another in a stratified pecking order. Each individual carried a full plate. When a person discovered that the detailed planning that had preceded the project had either underestimated or overestimated specific job requirements—and in a complex project like this it happened a lot—it was part of each person's job description to send this information up the line. Adjustment followed quickly. Full employment of all available resources was a project hallmark. A hundred and ten percent implied much more than an eight-hour day, but it also mandated sufficient R & R to ensure peak performance. Eighteen months into the project they were at least one month ahead of schedule, and there had been zero labor problems and no serious personnel accidents.

After a warm handshake with a clap to Alex's back, and a bear hug for Margo, Klaus led his visitors along the catwalk to the looming entry lock.

They exited the elevator-lock combination and walked down the carpeted tube, breathing the slightly compressed air. Sunlight flickered off curving walls, filtered through water and the clear composite openings that penetrated the tube at regular intervals. Shortly, they passed through the closed lock into Klaus' private chambers, and stood before his large window looking out over the frenzied activity throughout the underwater complex as it neared completion.

"I withdraw my comment about your view being unrivaled," Alex commented to Margo.

"Apples and oranges," she responded.

"Bratwurst and beer," Klaus added, easing the top off a bottle of Belgian micro. "They complement one another." He built a head and handed the mug of red-amber liquid to Margo. He tapped his Link pad before starting the second head.

The holographic display hovering over his table flickered, and then a bright red fire engine materialized at the left edge and swept directly through the middle of the display, shrinking as it moved, and ended in the upper-right corner, light flashing. Klaus grinned at the reactions of his guests.

"Fire wall protection," he said, handing the second drink to Alex. "But for this, that's not good enough either." He tapped his pad again and opened a third bottle.

What looked like a police squad car drove up next to the fire engine, and several uniformed police officers exited the vehicle and quickly positioned themselves all around the display. As Klaus lifted his froth-topped mug in salute to his friends, the holograph flashed, and a faint enclosure appeared around the entire display. For all the world, it looked like an old jailhouse. Alex's astonishment showed on his face. Margo registered her amusement at the playfulness this big man could display.

"Lock-up," Klaus muttered, slightly embarrassed, but also pleased at his friends' reactions. "To net security," he said, before quaffing from his mug. "My Link is completely isolated now," he added unnecessarily.

Turning serious, Klaus tapped his pad and made a few cryptic entries on his keyboard. His display, which had been showing a picture of the entire Western Complex, zoomed to a blur, ending with a translucent view of one of the turbines and its generator coupling.

"Look," he said, tapping his pad. The cursor changed to an arrow pointing at a coupling component. Another tap, and a part of that component filled the display. "Do you see it?"

Both heads shook.

Another tap. The display split, revealing components that attached to one another.

Margo gasped and Alex muttered, "Shit!" under his breath.

It was subtle, but displayed this way they both could see it. A shaft collar and its mount—when aligned to attach the collar to the mount, the holes didn't line up.

"Crazy thing is," Klaus said, "I checked the parts myself, and the holes do line up."

Alex shook his head in amazement. "First order tampering by someone who doesn't know jack shit," he said.

"But why? What's the point?" Margo asked.

"It doesn't make sense, does it?" Klaus added. "These are just informational drawings. My guys don't even use them for assembly." He took a swig of his beer.

"I agree," Alex said, "but someone's tampering. Some asshole hacked his way through all our normal safeguards and did…" He pointed to the split display, "this." He shook his head in amazement again. "It really doesn't make sense, does it?"

"So what do we do about it?" Klaus asked, setting his mug on the table.

"How many other changes have you found?" Margo asked.

"Actually, no more yet. I called you," he indicated Alex, "as soon as this turned up." He tapped his pad. "While you were coming I wrote a crawler program to look for more." The display shimmered. "It's a two-stager. First level compares with my personal downloads…"

Alex looked at him quizzically.

"Yeah, I know, but you know me by now," he said pointing to the roll of old fashioned drawings behind the couch. "I like to be sure." Klaus grinned at Alex. "Anyway, stage two looks for engineering mismatches using an algorithm I developed back at Darmstadt Poly. I activated it just before you arrived."

The display stabilized, and Klaus said, "Let's see what it found so far."

A list appeared that seemed one or two hundred items long. It grew as they watched. Klaus selected an entry and tapped his pad. An image flashed into the display. Klaus scrutinized it and announced, "Another mechanical mismatch."

Randomly, he selected several more entries. All of the modifications were elementary, and they seemed to relate to mechanical connections between moving parts.

"It's almost as if someone were trying to get the mechanisms to blow themselves apart," Margo remarked.

"But what kind of techno weenie would know how to do this," Klaus pointed to the display, "and not be smart enough to know that it doesn't matter?"

"Does it really 'doesn't matter'?" Alex asked, flipping through the displayed list, which had now grown to over five hundred items. "We can't just leave it, you know. We've got to put somebody on this…"

"Not a grunt, either," Margo interjected.

"Right…it's going to take time and money…" His voice trailed off as a look of concern crossed Margo's face. "Margo…?"

"The tube break you and I just investigated…If you were going to flood a tube, would you cut the top or the bottom?"

"So who says it was cut?" Alex wanted to know.

"You explain it then."

Klaus sipped his Belgian quietly, watching the discussion. Margo's slender figure silhouetted against the window was a disturbing distraction. The feelings she evoked in him barely reached the surface.

"Well, I…" Alex groped for an answer.

"Really, Alex, how do you explain a cut to the bottom of a tube suspended thirty meters below the ocean's surface?"

"You're sure it was a cut?" Alex asked the question seriously, although Margo was pretty certain that he already knew the answer.

"It sure as hell wasn't a rip, and a Great White would have severed the whole thing."

Klaus chuckled. "Don't need that problem," he said, "but you make a good point. Why go to all the trouble, and then cut the bottom?"

"Why hack into a secure system, and then change the wrong drawings?" Alex added.

Klaus chuckled again. "So we got ourselves a techno weenie with more weenie than techno."

"More balls than brains," Margo added.

"I'll buy the techno weenie part," Alex said, "but I want to think a while about the rest. We're missing something here." He stood up. "Each of you assign a computer nerd to the problem. You coordinate it, Klaus." He looked over at the "locked up" holographic display. "You understand this better than we do."

"The computer side, maybe," Klaus responded, "but a weenie didn't make that cut."

"We'll chase that one down," Alex said, as he and Margo headed for the lock. "Gotta go. Thanks for the beer."

"Hey, I'm your line handler, remember?"

Klaus got up and followed them through the lock.

CHAPTER FIVE

EASTERN COMPLEX—CIRCULAR DEFLECTOR

The sun had just popped its rim above the eastern horizon. The entire skydome was a cloudless gray, shifting to pale blue as Roxford Stewart looked over his eleven divers, already dressed in skins and gills, clustered at one end of a football-field-size barge, waiting for the action to commence. Dozens of sooty terns wheeled overhead, skreeghing a greeting to the morning sun, occasionally hitting the surface to catch a small fish. Several masked boobies were dissecting their early morning catch on the far end of the barge. A dozen or so lesser frigatebirds swooped through the terns, ambushing those carrying morning catches. One even attempted to rob the boobies at the other end of the barge. The sea rose and fell with a gentle meter--and-a half swell from the southwest, having virtually no effect on the heavily laden barge.

The barge bay behind the divers held sixty-seven concrete anchor blocks arrayed in five columns. The blocks were cubes, five meters on a side. Three massive meter-thick, pointed concrete pylons extended ten meters down from each block bottom. The blocks with their triple pylons stood balanced on their points like giant three-legged soldiers standing in rank at rigid attention. Each pylon was outfitted with three bronze paddle-like braces. The braces were two meters long and a half-meter wide. Once the anchor block and pylons were firmly embedded in the sea bottom, the braces were explosively extended, resulting in spikes that pointed upward at forty-five degrees. When extended, the braces acted like the barbs on fishhooks, ensuring that the 850-metric-ton anchors remained firmly attached to the sea bottom.

Stewart cast his gaze over the bird-filled sky, pausing to observe a mid-air clash, frigatebird on tern, pausing again as another frigatebird captured a piece of boobie breakfast. His eyes followed the columns of concrete anchors, swept past the extended arm of a massive crane, to land on his divers clustered on the flat deck area marking one end of the barge. Stewart was Klaus' lead diver on the Eastern Complex—big, strong, and very smart. Alex had brought Stewart aboard for his expertise and experience. A graduate of Duke University in mechanical engineering, Stewart had been hired by the National Oceanic and Atmospheric Administration (NOAA) to run its extensive diving operations along the East and Gulf Coasts. This included supervision of all shipboard diving activities as well. Following several years gaining a world of experience in all types of diving, Stewart accepted a position with GlobeTech, the worldwide oil rig contractor. Within a couple of years he headed their diving operations. Alex hired him away from GlobeTech with an offer he simply couldn't ignore. He was the lead diver in charge of the Eastern Complex diving activities, and—in fact—he controlled all the diving activities on the Jarvis end of the project. To be in charge of a major element of the largest, most complex engineering structure humans had ever attempted was beyond a dream come true for Stewart. To say he loved his job would be the understatement of the century.

The self-propelled barge was outfitted with station-keeping water jets. It sported a large crane at each end, specifically designed to lift and swing the anchors away from the barge, lower them to just under the surface, hold them while the divers attached the tensioners, and then release them for their plunge to the bottom five kilometers below. Both cranes were firing up.

"You know the drill," Stewart told his divers. "We got a significant surface current flowing east, moving with the swell, so you guys watch yourselves. If you lose it, you're going for a ride." He grinned. "No sense fighting it—you ain't that strong!"

Jeering and laughter passed through the ranks.

"I mean it, guys. If you find yourself heading toward Chile, just relax, and keep your EFCom activated. We'll pick you up with the chase boat as soon as we can get to you." He pointed at a swooping masked boobie. "And watch out for the boobies!"

General laughter.

"I want two guys in the chase boat, three at this end of the barge, three at the other end, and a stand-by at each end." He divided the divers into the two teams, and assigned Joey, a lanky former NOAA diver from Washington State, and Todd, a Texas oil rig diver, to the chase boat. "That leaves Gofer and me." He was referring to his diving supervisor Calvin Gofort, a Scottish import from the British offshore oil fields. "I'll take this end, and you, Gofer, take your team on a morning exercise." He pointed to the far end of the barge.

General laughter again.

"Let's move it, guys! The sun's awastin'. Let's see if we can drop ten today." And as an afterthought, "Remember the current!"

The barge maintained station to the east of the ribbon so that the prevailing current pushed it away rather than toward the ribbon. Station keeping was automatic, based on satellite GPS data. If it moved more than a specified amount in any direction, an alarm sounded on any number of individual Links. Whoever was charged with monitoring the barge's position would then take appropriate action. Two surface floats, supported by pontoons, rode on the ribbon side about a quarter of the way from each end of the barge to provide a platform for divers entering and exiting the water. Two accommodation ladders led down the five meters from the gunwale to the floats. As the barge emptied and it rose higher, the accommodation ladders would swing closer to vertical to accommodate the difference.

The divers knew better than to jump into the water, because of the strong current. They could have been outfitted with dive motors, but since they would remain shallow and virtually in contact with the anchor or ribbon all the time, there really was no need for the complication dive motors brought to a dive. They were bulky, needed recharging, and a diver had to keep track of it in the current when he was not moving. Stewart was happy to be without them on this operation.

Stewart watched Gofort lead his team on a trot to the other end of the barge. *That's one fucking great bunch of divers,* he thought as they assembled and headed down the accommodation ladder. Turning to his own team, he couldn't help but notice the sun glinting off Samantha's long tawny hair. "Sam" was never far from her partner, Woody, a tough, blond surfer from Southern California. They had both spent years looking for and finding lost wrecks off California, Florida, and in the Caribbean. Together they had pioneered rebreathers for recreational use. They adapted to gill technology without a hitch. They were naturals.

"Micky," Stewart said to his big farm boy from Nebraska who had joined the Navy to see the world, but discovered diving instead, "you got first stand-by."

"Aye, Skipper," Micky said. He loved nautical terminology.

"Sean," indicating the remaining diver, a big, tough Boston Irishman, "you and the love birds take the first drop." Stewart pointed to the accommodation ladder. Each team member grabbed a bag and hustled down to the float.

Quiet buzzing preceded the rubber-pontooned Zodiac as Joey brought it around the end of the barge, its flat bottom slapping the water. He brought it alongside the float, and Todd stepped to the float and held fast to the Zodiac.

"Hey Todd," Micky grinned at him. "Where's yer boots, cowboy?"

Todd gave him a stiffened middle finger.

"Where you want us, Boss?" Joey asked.

"Stay outboard of us til Gofer's guys get wet. Then hang out in the middle." Stewart looked over the equipment in the Zodiac—first-aid kit, folded KP, drinking water. "Joey, keep your EFComs on, and keep a sharp eye out. Todd, you got coms, okay?"

Todd jumped into the rubber-pontooned boat, and they sped across the open area to take station next to the floating tube.

"Comm check to Stu," Todd said over the EFCom.

"Roger."

"On station, Stu."

"Roger that."

"It's Stu; what's your status, Gofer?"

"Ready tae gie wet as suin as th' crane driver stabilizes th' anchur at three meters."

"Roger."

"Guid day Stu," it was Gofort again. "Who's drivin' th' cranes?"

"Whodoya think?" Stewart shot back. "We only got Doug and Tony over here."

"Hey Doog," Gofort said, "ye wi' me ur Stu?"

"I got yer back, Gofer," Doug answered. At 157 centimeters, Doug may have been the shortest crane operator anywhere, but he learned his trade at the Chicago docks, and there was none better.

"Ho Stu." It was Tony, the other crane operator. "You fuckin' ready to turn to?"

"Take it easy, Tony," Stewart said. "We're on open mike here."

"Sheeit!" That was Tony's answer to most comments, a hold over from his Brooklyn youth and his crane operations up and down the east coast.

"Okay, guys," Stewart said. "Tighten up the EFCom discipline. We got two simultaneous hazardous operations goin' on the same time. So cut the chatter!"

"Rogers" echoed across the bandwidth as each team focused on its individual tasks.

※

In addition to the divers and crane operators, two men were assigned to each crane, one to attach the lifting hook to the anchor, and a spotter to monitor the lift and communicate with the crane operator. They worked two dedicated spread-spectrum radio system codes that were not audible to the others.

The spread-spectrum radio system or SSRS—pronounced *sirs*—transmitter broke up the communication into tiny packets, and then distributed these packets across a broad band of frequencies in a specific pattern preselected in both the transmitter and receiver, similar to tuning an old-style transmitter and receiver to the same frequency. The resulting SSRS transmission sounded like a faint rush in a traditional receiver, but was crystal clear in the SSRS device. Since SSRS transmissions typically do not bend around the Earth's surface, transmission and reception were linked through GS32, high overhead. One side effect of SSRS was complete security of the transmission. Interception of such transmissions was a practical impossibility.

Tony swung his hook over the first anchor, and his hook man attached the massive hook. Before signaling completion, he checked to ensure that the tensioner reel was securely attached to the top of the anchor. Under the guidance of his spotter, Tony slowly lifted the 850-metric-ton anchor until the pylons cleared the gunwale. He swung the arm over the side and ran out the trolley. The barge tipped ever so slightly with the extended weight of the anchor so far outboard.

"You guys ready?" Tony asked as he prepared to lower the anchor.

"We're clear," Sean said. "Lower away—easy!"

"Keep comin'…"

"At the surface…"

"Easy now…"

"Submerged…Okay, hold it!"

As soon as Sean stopped the anchor, Woody and Sam, who were already in the water, darted to the tensioner mounted to the top of the anchor with a breakaway attachment. They hooked an inflated float bag to the attachment clamp, flipped the tensioner release, and pulled the end of the tensioner cable behind them as they dove under the nearby tube, aiming for the bright white vertical stripe that indicated the location of the attachment point and power socket on the bottom of the ribbon.

"Progress report." It was Stewart.

"Tensioner attached; power cable attached. Check continuity," Woody said.

Sam double-checked the connection while Stewart used his Link to verify continuity. Without it, it would not be possible to run the tensioner motor to pull the tube to proper depth, and then to hold it there. When fully operational, this tensioner motor along with all the other tensioner motors would maintain the two-meter tube exactly horizontal, to minimize any distortion of the vacuum sheath and its rapidly moving ribbon.

"Good continuity," Stewart said. "You guys clear out." And as an afterthought, "Watch out for the current."

"Barge seems to be blocking most of it," Sam reported.

Shortly thereafter, Sean reported, "We're clear. Let her go!"

"On my mark," Tony said from high above. "Five, four, three, two, one, mark!" He hit his release button.

One moment it was there, the next it was gone in a trail of bubbles, leaving the tensioner motor behind, suspended ten meters below the tube, unreeling the steel cable. Within five seconds the falling anchor reached a velocity of fifty meters a second. Thirty seconds later, moving at 330 meters a second, the 850-metric-ton mass smacked into the silted bottom 5,000 meters below the barge. Everybody waited, holding their breaths, waited…waited, and then nine sharp reports echoed from the deep, as explosive charges drove the bronze wedges into the surrounding silt.

The first anchor of the day was permanently married to the bottom.

A whoop and a shout emanated from the divers. Stewart tested the tensioner, pulling sufficient tension to dip the tube just below the surface, and then let it go back to neutral. "Tensioner functioning okay," he said. "It's your turn, Gofer."

※

Gofort's drop went without incident. Once it had been tested, Stewart gave the okay to move the barge 200 meters along the tube.

"You got it, Tony," Stewart said as he relinquished barge control to the crane operator high above.

The move took about ten minutes. When the barge returned to automatic station keeping, Stewart authorized both cranes to work independently for their next drops, under his and Gofort's supervision.

"Gofer," Stewart said, "let's do this from the barge deck." As he moved up the accommodation ladder, he directed the chase boat to move the floats right to the ends of the barge.

"You stand-by divers remain on the floats during the transfer," he said over the EFCom.

While the transfer took place he and Gofort reset the accommodation ladders. The entire transfer took only fifteen minutes, and shortly thereafter, the divers were ready to go again.

"Micky," Stewart said, "it's your turn to get wet. Sean, you got stand-by."

"BJ—ye an' Jessie switch noo," Gofort ordered at the other end of the barge, reassigning BJ, his "Joe Palooka" type Navy diver from

Minnesota to stand-by, and Jessie, a Philly Navy diver from stand-by to diving. "Maxie and Rod," Gofort said to his New Jersey oil rig diver, and his Georgia-born black Navy SEAL, "ye Laddies teach Jessie here a thin' or tois. An' be quick abit it."

Both teams worked quickly and efficiently without actually hurrying, each trying to produce the next set of explosions first. As it turned out, Gofort's team released its anchor a couple of seconds before Stewart's team. The resulting explosive charge echoes seemed to follow one another in perfect sequence—all eighteen of them.

As the divers passed high-fives, Stewart and Gofort sounded simultaneous alarms.

"Th' fekin' Blues are comin'!" Gofort shouted.

"Watch yourselves. You're surrounded by a bunch of Pacific Blues," Stewart said. "Joey, try to chase them off. Todd, get ready for in-water assist."

Joey cranked his engine to high, and tore through the school of sharks. The ones nearest the tube scattered, away and down. The inboard dozen or so headed toward the barge, stopped several meters away from the hull, and turned back toward the divers.

"Watch it, Lads," Gofort said, "theyur in attack mode," as several of the three-to-four-meter animals arched their backs, thrust their pectoral fins in a stiff downward angle, and began moving their forequarters left and right in a slow, jerky movement.

One of the four-meter creatures passed right by Gofort's float. In one smooth motion, BJ donned his mask, inflated his breathing bag, and leaped to the back of the Blue, left arm wrapped around the shark's head just behind its eyes. Before the shark could react, BJ plunged his knife into the shark's relatively soft underbelly, releasing torrents of blood, as the wounded shark twisted and turned, trying to dislodge its tormentor. The other Blues immediately turned their attention to the battle between them and the barge. One of the braver, smaller sharks launched an attack at the blood-spewing object. BJ twisted his entire body without letting go, and brought his victim's belly up into the face of the attacking shark. A moment later, it hit, disgorging shark guts into the surrounding water.

"Joey," Stewart ordered, "get the Zodiac between the frenzy and the other divers. Todd, boost them into the boat. Move it guys!"

"Okay, Lads—oot ay th' water reit noo!" Gofort ordered, sounding like a Scottish martinet. He ran down the accommodation ladder. "BJ, what th' fuk!" he said as he pulled BJ onto the float, a Pacific Blue nearly following him over the edge. "Ye crazy, Lad?"

BJ grinned at him as he wiped his blade on his skins. The water was a bloody boiling cauldron, shark dorsal fins flashing through the surface and disappearing underwater a moment later as the Pacific Blues competed with each other for their own flesh.

It was over in five minutes. The remaining two or three sharks disappeared into the depths, the current cleared out the blood, and it was as if it had never happened.

"Okay," Stewart said, "let's move the barge. You got it, Tony."

Joey brought the Zodiac alongside each float, dropping off the teams.

"Well, I'll be fucked three ways from Sunday!" Everybody immediately recognized Sam's voice. "Look at this, guys," Sam said as she held up one of her fins.

Half of the fin was missing, in the exact shape of a Blue's bite.

CHAPTER SIX

UNDERWAY BETWEEN BAKER AND JARVIS ISLANDS

Jarvis Island lay nearly 2,000 kilometers eastward and about 60 kilometers south of Klaus' pressurized quarters in the Western Complex. About 350 kilometers east of Jarvis, the final sections of the Eastern Complex were being prepared for submergence. Stewart and his divers had been dropping anchors for most of the previous month. The last ones were scheduled for that morning. Besides being a milestone, submerging the tube was too important and critical to the project for Alex, Margo, and Klaus not to be there. Besides, it gave them an opportunity to examine the entire length of the submerged rail, as they flew from Baker to Jarvis.

Klaus had arrived at the Baker Compound by skimmer just after sunrise, having pushed the surface-effect ship to its limits for nearly two hours. Alex met him at the dock, and they both had called on Margo as the sun snapped above the eastern horizon to peek through her living room window. Within minutes they were airborne, flying low over the line of buoys stretching 1,828 kilometers eastward. With throttles nearly full forward, they were crossing a buoy every twelve seconds, which was more than enough time to see if each buoy was in its proper position in the long line.

Margo, who was sitting in one of the back seats, reached into her shoulder bag, which occupied the empty seat beside her and extracted a thermos and three mugs. "Keep your eyes on the buoys," she said to Klaus, as the aroma of fresh coffee filled the cockpit.

"Black for the Cap'n." She handed the first steaming mug to Alex.

"White with two sugars." She wrinkled her nose as she handed the next to Klaus.

"*Ja*, just like my women—blond and sweet," Klaus responded, presenting her with his wide open grin.

Margo dropped her eyes as she poured her own cup, adding a bit of cream to smooth out the coffee's edge. They flew on in silence, buoys flashing beneath them monotonously. They sipped coffee, enjoying the ability to relax without guilt. The day had started early, especially for Klaus, and the buoys were where they were supposed to be. The observations were routine. If anything were out of the ordinary, it would stand out like a sore thumb. Margo made sure that at least one of them had the "buoy watch." This way the other two could doze, she commented, and then corrected herself, suggesting playfully that Alex "Keep his mind on his driving, and his hands on the wheel."

BUOY 1528

Three hours and fifty minutes into the flight, about three hundred kilometers out of Jarvis, Alex had the Buoy Watch, and was scanning ahead, looking for the next buoy. Something caught his eye, and he instinctively pulled back on the stick and banked right, up and away to the south. The sudden change in aspect jolted Margo and Klaus to full wakefulness.

"What is it?" Margo asked from the back seat.

"I don't know yet," Alex answered. He was not sure what he had seen, but it looked for all the world like the buoy ahead had grown to several times its normal size. *But it can't be*, he thought, scanning to his right as the aircraft completed a full turn. *It has to be a vessel of some kind. Sonofabitch…have we finally found them?*

"Something down there," he said aloud, pointing to the buoy ahead. They were now 300 meters higher and throttled back to bare headway.

Margo and Klaus saw it at the same time.

"That's a small two-masted schooner," Klaus said, reaching for his binoculars.

"Marconi-rigged, sails furled," Margo added, needlessly.

Klaus focused on the vessel and remarked, "Looks like it's tied to the buoy."

They're going nowhere fast, Alex decided, dropping low to circle the vessel. His certainty grew as he approached.

"I think they have divers in the water," Margo said as they completed the first circle. "Look at the gear on deck…"

"And that's a stand-by diver," Klaus added pointing to a person on a chair wearing a shorty wet suit and conventional compressed-air scuba bottles.

Gotcha, you bastard, whoever you are, Alex thought. "I think we finally may have some answers," he said aloud, and activated his SSRS throat mike.

"Jarvis, this is *Floater One*."

"This is Jarvis. Go!"

"This is Alex on *Floater One*. I'm circling," Alex checked his chart, "buoy one-five-two-eight. We have an unknown sailing vessel moored to the buoy. I say again, we have an unknown sailing vessel moored to buoy one-five-two-eight."

The traditional formality of radio communication protocol was rendered unnecessary by the introduction of SSRS. Alex ran a relaxed radio protocol. So long as everyone understood what was happening, he figured, there was no need to complicate things with unnecessary language.

"Roger, an unknown sailing vessel moored to buoy one-five-two-eight."

"There are divers in the water. I'll land to investigate. I have Margo and Klaus with me." He consulted briefly with the other two. "Margo and Klaus will conduct a wet investigation. Dispatch immediate skimmer assistance."

"Roger, you're landing to investigate. Roger, the presence of Margo and Klaus. Roger, their wet investigation…" There was a pause of a few seconds. "Roger, the skimmer dispatch." Another pause. "Expect *Skimmer Two* arrival in one and a quarter hours, Alex."

"Is *Floater Two* available?"

"Wait one…" And after a longer pause, "Roger that. *Floater Two* is serviced and standing by on Jarvis."

"Very well. Dispatch *Floater Two* with three divers…make that three *big* divers!"

"Roger, Wilco…" and after several seconds, "ETA *Floater Two* is forty-five minutes. What about *Skimmer Two?*"

"Dispatch both *Skimmer Two* and *Floater Two*. Continue monitoring this channel." Alex deactivated the SSRS transmitter, and turned to Klaus in the seat beside him. "Are you guys really okay with this?"

"I'd prefer a couple of my tough boys," Klaus said, turning to look at Margo, "but Margo's a better diver than all of us put together." He winked at her. "Anyway, in the water strength is much less important than skill and agility." He reached back and squeezed her hand. "Besides, I'm big enough for any two other guys anyway."

Alex grinned at this response. *But I should be going down with her*, he thought, remembering the recent incident with the sharks. *I protect what is mine…* His thoughts tumbled over one another as he circled lower. *Mine—she's not mine… is she? Is she anybody's? Who does she sleep with?* He straightened out his glide path and pulled back the throttles as he flared over the low waves. *Not you, stupid!* The plane touched the wave tops and settled down into the water as Alex throttled back completely. *Klaus…?* Alex put the thought out of his head as he taxied toward the moored schooner.

ON THE SURFACE AT BUOY 1528

Onboard the vessel, people hurried about frantically, obviously in a panic at the plane's arrival. Alex could see no clear purpose to their scurrying about, almost as if they were without direction. Several were looking over the side.

As the plane bobbed gently in the light swell, Alex said, "There are two sets of skins, gills, and other gear in the luggage compartment. The largest one is blue." He grinned at Klaus. "Sorry, guy, but it'll stretch.

He scanned the schooner with his binoculars. Painted prominently in vivid green letters, the vessel's name stood out in sharp relief: *Green Avenger*. Aboard the vessel there appeared to be some preparation for getting underway, but to Alex it seemed particularly inept. Most of the crew looked young, in their late teens or early twenties.

Beside him, Klaus struggled to get out of his clothes and into his skins. His size made a comical picture. Alex chuckled without comment. He looked over his shoulder, but quickly turned back to the front, his mind filled with the picture of barely covered lithe loveliness. She seemed not to have noticed his glance, and Alex felt an old familiar pain knot inside his chest. How many women he'd known had ever possessed the capacity to understand him? This one—he resisted the urge to turn around again—did, in spades.

※

Margo had just retrieved the diving gear from the luggage compartment behind her. She handed the larger blue skins to Klaus who had been struggling for several minutes in the confined space of the cockpit to get his clothing off. With the twin luxury of having more space and being especially lithe and flexible, Margo easily slipped out of her shorts and top. Sitting there in the flimsiest of bra and panties, she felt a flush arise from deep in her core. As she had slipped off her top, from the corner of her eye she had seen Alex turn. Now she watched with amusement his struggle to keep his eyes forward. From her vantage point, Klaus was all skin. She glanced from one to the other and let her fingers brush across her lace-covered nipples. Her mind started to drift, and she angrily pulled herself back to the present situation, and slipped into her borrowed skins.

As she sealed the front of the black-with-red-trim skins across her breasts, Klaus encased his legs and proceeded to work his arms into the tight sleeves. Margo reached over the seat back and stretched a sleeve to facilitate his task.

"Push," she ordered, pulling the narrow wrist over his large hand. She noticed the tremendous difference between his hand size and hers. She was big for a woman, and had always considered her hands large and ungainly, but next to his, they seemed small and dainty. The feeling was strange and quite unaccustomed.

Margo let her hand trail across Klaus' large chest as she reached for his other sleeve. As she pulled the narrow opening over his wrist, she noticed Alex looking at her tenderly, and her heart did a flip-flop. She was on the verge of melting, and she angrily took charge of her emotions.

"We'll finish dressing on the pontoon," she said, slipping her fins over her arm. She opened the right-side door. With that action, she regained control of the situation and her emotions, and slipped down to sit on the pontoon, her feet dangling in the warm tropical water.

The front door opened, and Klaus handed out her gills and mask. She donned the vest-like contraption with its tiny flask of ultra-high compressed air, and then accepted Klaus' fins and gills. Klaus emerged from the cockpit and eased down beside her on the pontoon. She handed him the gills, and he sighed as he slipped his arms through the holes.

"Size is not always an advantage," he remarked, sealing his gills across his chest.

Margo handed him the fins.

Above them, Alex fumbled in the back seat, retrieved two shark sticks, and handed them out the door. Klaus accepted them and handed one to Margo. She checked the mechanism, extending and collapsing the telescopic tube before attaching the device to her utility belt. Her mind flashed back to her last dive with Alex, and she missed George—a lot. She looked at Klaus, who somehow had the ability to make the gill unit across his back and chest appear small. She took her console and lifted it, pointing to a blank spot with her right hand.

"I miss my dolphin call," she shouted above the noise of the feathered propeller, meaning it.

"What do you propose?" Klaus shouted back, deferring to Margo's greater expertise. He scanned the water in front of them, his gaze resting on the moored vessel.

Above them, Alex cut power to the right engine. It whirred to a stop, the three-bladed prop snapping to a standstill. The light breeze carried the noise of the other engine away from them, so they were surrounded by relative quiet. Then he opened the door and tossed a small buoy into the water in front of them.

"Insurance," he said.

Margo reached out to grasp the buoy tether, and fastened it to a cleat on the pontoon. A thin antenna extended about thirty centimeters above the buoy; it was similar to one extending from the buoy a hundred meters away. "We don't really know what they're about," she said, "but unless they're in distress, we can order them away from the buoy."

"But the divers."

The way Klaus said the words, clipped with his teeth clenched, told Margo the extent of his anger. She remembered his intensity when he showed them the computer system tampering. He was jumping to conclusions here, but, she conceded, it was an obvious leap.

"Assume the worst," she felt her own anger rising. "Assume these guys have been slicing the tube undersurface." She pounded on the pontoon with her fist. Alex stuck his head out the still open door.

"What?" he inquired.

"Are you carrying a KP?" She was referring to the Kevlar portable recompression chambers used for emergency treatment of divers suffering from decompression sickness. Alex insisted they be available at all diving sites on the Project.

He pointed down. "One single, and two more on the way."

Margo looked between her legs and realized she was sitting on one, tightly folded and strapped to the pontoon surface.

"Margo," Klaus said, pointing at the KP on which they were sitting. And then he outlined his proposed plan of attack.

"Are you sure this is a good idea?" Margo asked, her voice expressing concern.

"We don't know anything about them—nothing. How many are in the water? Are they armed? With what? How will they react to our presence?" Klaus gave her a crooked grin, raising an eyebrow. "It's your bailiwick, I know, but do we really want to take any chances here?"

After considering it for a few moments, Margo nodded her acquiescence and pointed to Alex sitting in the cockpit door. Klaus stood and spoke with Alex who also nodded his agreement. Then Margo set her bone speakerphone and throat mike, and donned the completely transparent full facemask with an internal cup over her nose and mouth, adjusting its seal so it would maintain a slightly positive internal pressure.

"Let's do it!" she shouted, stepping back into the clear, blue water, readjusting her mask as she slipped under the surface, letting the enveloping liquid wash away her lingering emotions. This was business; she was in her element, as good as any, and better than most.

Klaus grinned while waving to Alex, adjusted his mask, and rolled into the water to join her.

CHAPTER SEVEN

SUBMERGED ON GILLS AT BUOY 1528

Visibility simply doesn't get any better than this, Margo thought as she adjusted her buoyancy about five meters below the surface. The water seeping into her skins was close to body temperature; she could have spent several hours in this water without any protection at all. It was so quiet she could hear the soft whir of her gill pump. She breathed deeply, imagining that the oxygen in the air she inhaled somehow tasted better than that from old-fashioned scuba tanks.

Klaus entered her peripheral view, settling into neutral buoyancy alongside her. By unspoken agreement she was in charge down here. Margo fingered her console, and a faint, rhythmic rushing filled her head. She signaled to Klaus, and he activated his EFCom unit.

"EFCom test…one, two…" Margo counted slowly while Klaus adjusted his reception volume by tapping the back of his console.

"Test back… one, two…" Margo heard over the rhythmic rushing caused by Klaus' breathing. She tapped her console until the computer filters eliminated his breathing noise, leaving only the crystal clear sound of Klaus' voice inside her head, like an intimate, caressing whisper, courtesy of her bone conduction speakerphone.

"Base test…one, two…" Alex tested into the circuit.

"Clear," Margo said.

"Clear," she heard Klaus' answer.

"Let's go, Klaus!" Margo's voice was picked up by her throat mike, where it changed to electrical signals that were digitized in the Electrostatic Field Comm transceiver integrated into her gill pack. The resulting digital packets were exactly the same as those produced by a SSRS transceiver. They left her short antenna as an electrostatic field, and for the most part were absorbed by the surrounding water. But a small portion of the energy reached the surface where it spread out concentrically from a point directly above her, captured in a millimeter-thick surface layer. At light speed the packets arrived first at the back-up buoy, and a meaningless moment later at buoy 1,528. In both cases they were processed and retransmitted into the air and back into the water with significantly increased power, enough to reach a diver as deep as one hundred meters or a kilometer distant horizontally. By appropriately keying their consoles, two or more divers could limit their conversations to any combination of participants, in or out of the water, close by or far away. In effect, the marriage of EFCom and SSRS eliminated the isolation that once was part and parcel of diving.

"Vector us in," Margo said to Alex.

"Three four seven," he answered, indicating that buoy 1,528 lay to the north and slightly west of their present position.

Margo and Klaus each set their consoles, and a faint line appeared down the center of each faceplate. Off to the right side a glowing red spot indicated their present orientation with respect to the set vector. As Margo turned to her left, the red spot aligned with the faint line, and she kicked off.

"Let's go, Klaus!"

Margo angled down until she reached ten meters, all the while kicking with a strong measured stroke. Klaus followed closely. Soon they saw the shadow and then the hull of the small schooner, and below it the suspended tube disappearing in the haze to the left and right at a depth of thirty meters.

"There, Klaus, look!" Margo pointed to the suspension cable.

Three divers clung to the cable, descending while they watched: a male with loud, floppy swim shorts, longish brown hair, and a full short beard; a bikini-clad shapely female with long dark hair forming a halo around her head; and an exceptionally thin male with yellow hair floating around his head that looked like it would settle into a Dutch page-boy cut when dry. Shimmering silvery bubbles cascaded to the surface above them.

"Regular scuba," Klaus remarked unnecessarily.

Margo and Klaus reached the cable surrounded by the rising bubbles from the scuba divers below them.

"I don't think they know we're here," Margo said, removing her shark stick and extending it to its full length. "Let's give them a surprise, shall we?"

"Okay," Klaus answered, removing his own stick as Margo moved several meters away from the cable. Each released a bubble as they began to descend fairly rapidly. The scuba divers below them continued their own leisurely descent, still unaware of the action over their heads.

A meter above the scuba divers, Klaus slowed his descent and placed himself horizontally in the water between them and the in-streaming sunlight. Off to the side and somewhat above him, Margo could see the shadow he cast. As the water above them darkened, the three scuba divers stopped and looked upward. Their reaction to Klaus' imposing figure silhouetted against the sun-drenched surface was quite comical. The bearded male and the female ducked and began to swim awkwardly away from the cable in opposite directions. The blond pageboy seemed frozen at the cable. After a few moments the swimmers stopped and looked up again.

Klaus chuckled, audible only to Margo and Alex, of course, and waved to the divers with his left hand while pointing at them with the shark stick in his right. Margo moved in and placed herself above the nearest scuba diver, gesturing with her stick.

Unable to communicate with each other except on the most basic level, the two divers gesticulated frantically, and cast about to orient themselves, having moved away from the "safety" of the cable. One of the divers pointed up, and the other nodded in exaggerated affirmation.

"Now!" Margo said. Both she and Klaus dropped a meter. Carefully avoiding the firing button, each forcefully jabbed the scuba diver below, aiming for the buttocks area.

Reacting to the painful poke, each scuba diver stopped and rolled over, facing the looming menace above. Margo's female diver reached out attempting to grab the end of the shark stick.

"Careful!" Klaus warned. "You don't want to lose it!"

"On the other hand," Margo answered, "a little demonstration couldn't hurt."

"Talk to me!" Alex's voice surprised both of them. "What's happening?"

Margo briefly described their situation while dodging the attempts of the diver below her to grasp her stick. They both drifted upward.

"Watch your depth!" Klaus warned, poking at his own diver who was trying to imitate Margo's opponent.

"Do you have spare charges?" Alex asked.

"One," Margo answered.

"Two," Klaus added.

"Why not?" Alex said with a chuckle in his voice. "Those things are pretty impressive, you know."

"Do it, Klaus," Margo ordered, pulling her stick away from her diver's grasp.

A moment later a bright flash followed immediately by a loud, sharp crack filled the water surrounding the five divers. The end of Klaus' stick seemed to leap out toward the diver below him, as a stream of carbon dioxide bubbles burst against the scuba diver's chest, throwing him back and down more than a meter. Klaus fumbled with his stick handle, and in a moment a tumbling small metal cylinder flashed into the depths. The terrified scuba diver discovered that he was still alive and unhurt just as Klaus slipped a new cartridge into the stick handle.

"Wheee…!" Margo shouted, waving the end of her stick toward her diver, who abruptly ceased all aggressive behavior. Klaus' diver appeared equally contrite. Klaus turned toward the third diver still on the cable who lifted his hands toward Klaus, palms outward.

"Okay," Margo said, "let's drive them down." She waved her stick at her diver, gesturing downward with her left hand. Klaus followed suit, and then pointed his stick at the third diver. Soon all five divers had passed the tube and were hovering twenty meters below it near the fifty-meter mark. Klaus herded the scuba divers directly below Margo and handed his stick to her, glancing at his console as he did.

"Watch your profile," he cautioned as he rose in the water column directly above them. He stopped his ascent forty meters above Margo and her three captives, and hung there, watching the scenario play itself out.

The scuba divers nervously checked their own consoles, and then pointed to their tank gauges and dive timers with exaggerated gestures. The female made a move to ascend, but stopped immediately when Margo waved one of her sticks toward her.

It took less than fifteen minutes. Margo filled the time by describing the divers' reactions to Klaus and Alex. Finally, the bearded male pointed to his mouthpiece, eyes wide with fright. Margo waved a stick in that diver's direction. The diver responded by slashing a hand across his throat several times, a universally understood gesture meaning "I am out of air!" By this time the diver was clearly in a panic.

"Here they come," Margo said, as she tucked the second stick under her arm and pointed to the surface fifty meters above them. "How's your profile, Klaus?" she asked.

"Fine. I can surface with them."

The three scuba divers shot past Margo, surrounded by a rising cloud of bubbles. They were well past normal decompression limits, and they knew it. Nevertheless, they obviously had chosen the risk of decompression sickness over that of drowning.

"Beginning my ascent," Margo said, holding her console in front of her. "It's going to take a while."

ON THE SURFACE AT BUOY 1528

Klaus inflated his bag as he reached the surface, and turned to the thrashing divers between him and the schooner.

"Alex, activate the loud speaker!"

"You scuba divers," he said, his voice booming across the water from the floatplane. "Inflate your BCs and stop thrashing around!"

Moments later the three terrified divers ceased their struggling as their air-filled buoyancy compensators took over the task of keeping them floating on the surface. Klaus approached them.

"Speaker off," Alex said in his head.

Klaus looked down to monitor Margo's progress. "How we doing, Margo?"

She waved from twenty meters. "I'm on profile. In eight minutes I'll switch to pure oh-two at ten meters. Watch those guys, Klaus. They're going to be in serious trouble any time now."

"Alex…what's *Floater Two's* ETA?" Klaus asked as he approached the scuba divers.

"Fifteen minutes."

"What about *Skimmer Two*?"

"About a half-hour, Klaus."

On the surface, the female diver threw off her facemask. "Oh God, it hurts!" She looked around in a panic. "Help me!" she screamed, spotting Klaus several yards away.

"Alex, switch *Floater Two* into the circuit. We got a problem here."

"Roger that—*Floater Two*, you are now on line with Klaus and Margo."

"Klaus…it's Tex. What's your situation?"

Klaus briefly explained the nature of the problem to Carlos Sanchez, his lead pilot at the Eastern Complex, whom everybody called Tex. They decided to drop two divers and the Zodiac near Klaus, and then land *Floater Two* and get a third diver and each of the KPs from the two planes aboard the vessel.

"I'm coming in, Klaus. Wave at me!"

As Klaus waved, the sleek two-engine amphibian aircraft passed low overhead, and the inflatable Zodiac splashed nearby. On its second pass two divers hit the water; one went for the Zodiac, the other toward Klaus.

"It's Stu, Klaus. Jonesy over there is a new arrival." They were EFCom equipped. "We were just suiting up for an inspection dive when Dispatch called. I figured Jonesy might as well get his feet wet."

"I'll get the KP from *Floater One*." From his accent, Fred Jones clearly had his roots in the northeast, possibly Philadelphia.

※

Sanchez brought *Floater Two* to a smooth stop near *Green Avenger* just as Jones arrived with the Zodiac, propelled by a tiny electric motor. Gofort, the third diver, clambered out of the plane and loosened the KP attached to the right float. As Jones came alongside, he rolled it into the boat and followed himself.

"Okay, Jonesy," Gofort said, with distinctly Scottish overtones, "Let's gie th' air unit from th' other float." Since all the participants, in or out of the water, were interconnected with their EFCom units, everyone remained aware of each part of the operation.

Moments later the Zodiac arrived alongside *Green Avenger*. Jones shouted instructions to the crewmembers about two meters above him, and a line snaked down, followed by a rope ladder they attached to a cleat at the gunwale. Gofort grabbed the top rung of the flexible ladder and hoisted himself to the schooner deck. Jones attached the end of the line to the first KP package and signaled to Gofort. Gofort handed the end of the line to two crewmembers and told them to walk it to the other side of the schooner. As they complied, the fifty-kilogram KP reached the gunwale, and Gofort guided it over and onto the deck. He then tossed the end back down to Jones, and they hoisted the second KP to the deck. Then Gofort tossed the line back to Jones again, and while Jones attached the air unit, Gofort laid out the KP chambers side by side on the deck. They looked like shriveled-up long green balloons. The guys repeated the performance to bring up the air unit, and moments later Jones clambered over the safety line.

With professional proficiency, the two divers quickly attached hoses and hooked up the fuel cell generator to run the compressor.

※

In the meantime, Klaus and Stewart herded the three panicked divers to the Zodiac alongside the schooner. Klaus pitched himself up and over the inflated tube that made up the side of the Zodiac. Then he crossed his hands over the side and grabbed the female diver by her outstretched hands. Without letting go of her hands, he uncrossed his hands and pulled her into the Zodiac with one deft maneuver that flipped her on her back and tanks as she came out of the water. *She couldn't be more than eighteen or twenty*, he thought as he stripped her tanks and fins, hooked her BC to the line from the schooner, and signaled Gofort to haul her up.

While she was in mid-air, Stewart called out, "What's your name?"

"Tiffany, Tiffany Montague," she said with a pain-filled grimace, while she attempted to grab the rope ladder.

"Don't do that, Tiffany," Stewart said. "Let us do the work."

As the guys assisted her over the safety line, the bearded diver yelled out, "I'm hit, I'm bent. Help me, please!"

Klaus repeated his flipping performance on the bearded diver who said he was Carmine Endsley, but people called him Carey. He appeared to be in his early twenties.

※

Simultaneously up on deck, Gofort quickly briefed the ailing Tiffany, outfitted her with a throat mike and a bone headset, and clipped a miniature SSRS unit to her bikini halter. The KP SSRS units were set to a special code that only the KP operator could receive, to keep the recompression chatter off the whole circuit. As Jones tucked Tiffany into a KP, explaining to her that it was like snuggling into a sleeping bag with a window, Carey arrived on deck clutching his left elbow and wincing with pain as he stepped forward and limped toward the other KP.

While Jones quickly briefed Carey and outfitted him with another SSRS unit, Gofort started the air unit and began pressing Tiffany down to fifteen meters. Through the window he could see her pinch her nose to clear her ears, her eyes still filled with fear. A minute later, Carey was looking through the clear window of the second KP, his eyes as fearful as Tiffany's, as Gofort pressed him to fifteen meters.

"Ye troaps okay?" Gofort asked after both arrived at fifteen meters.

"I'm fine," Carey answered. "The pain's gone."

"Mine too," Tiffany added. "How long will this take?"

"Several hoors," Gofort said. "'S'better than dyin'."

※

In the Zodiac Klaus had his hands full with the third diver who appeared to be asymptomatic. Klaus had pulled him into the Zodiac and helped him with his gear. Almost immediately, the diver launched into a tirade.

"You nearly killed us. This is piracy. Who the hell do you think you are?" He demanded to be put back on his vessel.

Klaus reported the gist of the one-sided conversation to Alex.

"Does he have any decompression sickness symptoms?" Alex asked.

"Not that I can see, Alex. At least not yet. When's *Skimmer Two* arriving?"

"Wait one," Alex said. "*Skimmer Two*, you copy?"

"*Skimmer Two*, aye." The voice carried an unmistakable Jamaican lilt. "I be there 'bout twenty minutes. I got another KP with me."

Klaus looked at his charge. "What's your name?" he asked.

"Lars Watson," the diver answered, and then he suddenly doubled over in pain. "Oh shit," he moaned, "my left arm is on fire, and my right leg is numb, you bastards."

Klaus reported the problem to Alex.

"Margo," Alex said, "are you surfaced yet?"

"On my way to the surface right now," Margo answered, and then broke the surface next to the Zodiac.

"Klaus, get this jerk on my floater. Then join Margo and both of you stand by on the schooner," Alex said on the circuit. "Stu, you join Tex in *Floater Two*. Tex, head on back to Jarvis. And Stu," Alex added, "get the recompression system ready."

Everybody acknowledged and set about carrying out his orders.

"Gofer, you join me in *Floater One*." He paused. "*Skimmer Two*, what's your heading?"

Skimmer Two told him, and Alex said he would fly a reciprocal heading and meet him in ten minutes.

AIRBORNE TO JARVIS ISLAND

In five minutes Alex was airborne with Gofort and Watson just ten meters above the wave tops to minimize Watson's symptoms. He set maximum throttle, heading toward *Skimmer Two*.

Eight minutes later Alex triggered his SSRS. "I've got you visual, *Skimmer Two*. Set her down. I'll come in beside you."

"*Skimmer Two*, aye."

As the surface-effect ship settled into the water, Alex landed his amphib and came up alongside *Skimmer Two*. By this time Watson was beyond talk, doubled over with pain, and terrified that he would die.

Jefferson Carver, *Skimmer Two* skipper, wiped beaded perspiration from his nearly blue-black hairless pate and round face with a large kerchief, and handed Gofort a line to secure the two vessels together.

"Haya Gents. How the laddie be doin'?" he asked, his black face split by a wide grin that belied his top notch rating as a skimmer pilot.

"Th' laddy," Gofort said with dripping Scottish sarcasm, "was sabotagin' th' tube when we caught heem."

"Is dat so, now? We'll bloody well have to do sumpin' 'bout dat," Carver said, as he single-handedly reached out and hauled Watson to the skimmer deck.

Gofort joined him, and between the two of them, they stretched Watson out, and stuffed him unceremoniously into the KP after rigging him with a SSRS unit. As the KP pressure increased, Watson's face through the window began to relax, and by the time he reached fifteen meters he was pouring out additional vindictive at his "tormentors," as he called them.

"Dat lad got a mouth don' quit," Carver quipped as Gofort tossed the line back to Alex.

In five minutes Alex was airborne again by himself, headed back toward the *Green Avenger* with *Skimmer Two* following in the water at a slightly slower pace.

"What's your progress, Klaus?" Alex queried by SSRS.

"The schooner's ready to get underway for Jarvis, but it'll take about thirty hours," Klaus said.

"We'll be there in ten minutes," Alex said. "Offload the KPs to *Skimmer Two*, and put Jonesy on the schooner." Alex paused, working out the details. "Klaus, you and Margo join me. *Skimmer Two* will get the patients to Jarvis ASAP; Stu will have the Jarvis recompression system ready to receive them." As an afterthought he added, "Issue Jonesy a sidearm and have him bring the *Green Avenger* to Jarvis."

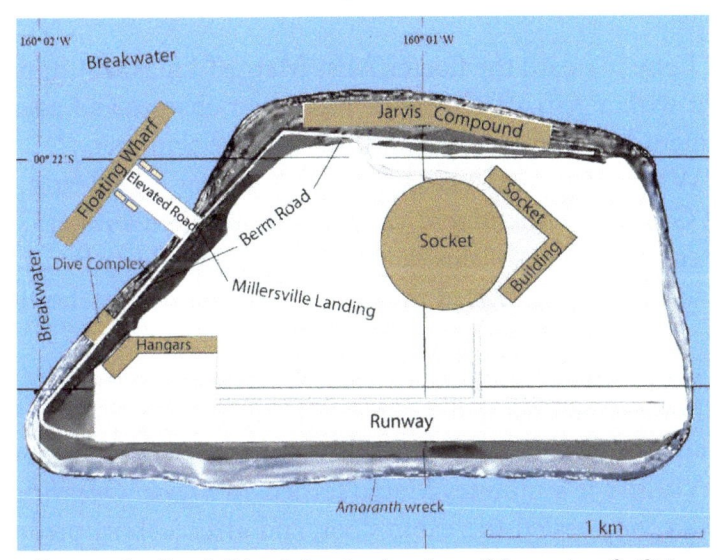

Imagere 5—Jarvis Island showing locations of the various facilities

CHAPTER EIGHT

ON THE SURFACE AT BUOY 1528

Margo watched as *Skimmer Two* settled into the water beside the clean lines of *Green Avenger*. She couldn't help but reflect on the stark contrast between the 200-knot futuristic skimmer and the classic two-masted schooner beneath her feet. As with so many things in her life, she was torn between two seemingly unattainable things. The speed and high-tech nature of the skimmer were awesome, but the utilitarian lines of a hull designed to move through the water with maximum efficiency propelled only by the wind—how could she not love that? Both were as modern as tomorrow, but somehow, that schooner.... It was a bit like the difference between Alex—extraordinary engineer surfing the leading edge of the technology wave, and Klaus—classic curmudgeon who kept up with Alex in everything that mattered.

Margo shook her head in annoyance. It always came down to that. *Thank goodness mindreading is just a myth,* she thought as she watched Klaus assist the other guys transfer the two KPs to the skimmer.

Alex had made a couple of turns in the air above them waiting for the activity to slacken. Then he landed and taxied up to *Skimmer Two*.

"Time to board the floater, Miss Margo," Jones said, approaching her with a grin after the KPs with their charges and associated equipment had been safely stowed aboard the skimmer.

"What's that for?" Margo asked, looking at his sidearm, a holstered Glock with its fifteen-round magazine filled with .357 loads.

"I'm the master of this domain for the next thirty hours or so," Jones said. "This," he patted the Glock, "is to make sure the children behave."

"Do you really think you'll need it?" Margo asked.

"No, but the Boss said to carry it." He grinned again. "Just having it should assure peace and tranquility."

Within a few minutes Margo settled on *Floater One* with Klaus and Alex. She sealed the cabin door, and after a short preliminary run, *Floater One* lifted into the air as Alex set a course for the Eastern Complex. They were running over two hours behind schedule, but when Margo pondered all that had happened in that short time she shook her head in amazement. Once again she was struck with Alex's ability to lead men in difficult situations. The professionalism of his crew was nothing short of amazing. She knew their day had been planned, and yet they quickly adjusted to this emergency as if it had been part of their plans all along. That was, she knew, part of their professionalism, but without Alex it wouldn't have happened. And with both Alex and Klaus working together…

"Tell the press we're on our way," Margo heard Alex transmit to the Eastern Complex. "We're about one and three-quarters out."

Sitting in the back seat of the floater, Margo felt that familiar warmth rise in her again as she glanced from Alex's full head of hair to Klaus' polished pate. *They're so different*, she thought, suppressing any further feelings. *I am a professional engineer*, she said to herself. *I will not let emotions rule me!*

ABOARD *GREEN AVENGER* UNDERWAY FOR JARVIS ISLAND

"Okay, boys and girls," Jones announced in his Philly twang to the assembled motley crew of *Green Avenger*, "I'm taking *Green Avenger* to Jarvis Island. We got thirty hours ahead of us. I can do it by myself, but I would rather do it with a bit of help, so this is how it is.

You assist me voluntarily, or I tie you up and lock you below. Since I will be pretty busy, there won't be time for toilet or food breaks." Jones paused, waiting quietly.

One youngster stepped forward. "I'd be honored to assist. We're not violent people." He spread his hands in front of him. "After all, we only want to protect the environment."

Heads nodded around the group.

"No trouble from us, Sir," from a cute teenage girl who seemed barely old enough to leave home, let alone be on a small schooner in the middle of the Pacific.

Jones shook his head in dismay. "How old are you, Kid?"

"Seventeen," she answered, "but I'm serious about the Environment."

Jones could clearly hear the capital "E" in Environment.

"What's your name?" he asked her, trying hard not to stare at her short bleached hair replete with brilliant stripes of green and purple, and her pierced cheek, lips, eyebrows, and tongue.

"Francesca," she answered impishly. "What's yours?" She stood with legs apart, hands on hips, completely oblivious to the trouble she and her friends were in.

It was going to be a long thirty hours.

JARVIS ISLAND RECOMPRESSION COMPLEX

Forty-five minutes later *Floater Two* taxied to a stop on the apron at the west end of the brand-spanking new runway that paralleled Jarvis Island's south beach. Stewart headed off trailing Sanchez to prepare the recompression chamber for the arriving guests. Stewart's regular operating base was the Eastern Complex, but given the time element in this bends situation, Alex clearly had concluded that using the Jarvis recompression facility was the best choice for the patients. In his career, Stewart had been there before, like the time three of his divers got hit on a rig under construction 500 kilometers off the Louisiana coast. Their chamber wasn't yet functional, but a Coast Guard Cutter eighty kilometers away had an operational chamber that could hold three patients in an emergency. They helped, and Stewart saved his divers. Or the time a scientist diver in saturation mode surfaced off St. Croix instead of returning to the habitat on the sea floor—Stewart opted to use a local chamber in Christiansted instead of waiting for the NOAA research ship to arrive with its more sophisticated chamber complex.

While Stewart was readying the recompression system, *Skimmer Two* announced it would deliver the patients in about twenty minutes. Stewart wasn't looking forward to this task. These creeps had attempted sabotage of *Slingshot*. Stewart's approach to problems like this resulted from his conditioning on isolated oil rigs in remote parts of the planet, where there was little time for the civilized niceties of lawyers and trial. In his world bad people had accidents. Fixing these guys did not sit well with him, but one thing he had learned during his months with *Slingshot*—the boss had impeccable judgment in these matters. So he would do his best to get these idiots back on their feet.

The Jarvis recompression system was not very sophisticated. It consisted of a low coral-white building containing a single two-lock chamber large enough to hold four or five sitting people, but barely large enough to hold the three KPs and an attendant. It had a bank of four permanent compressed-air cylinders with the ability to add any number of portable cylinders on a cascaded manifold. It had a main compressor and an emergency backup. It also had an oxygen manifold served by an expandable bank of oxygen cylinders. There was no oxygen-generating capability or mixed gas capability, except for the old-fashioned method of mixing the contents of cylinders containing known amounts of specific gasses. For what the Project needed, however, it was a perfectly fine backup facility to the amazingly state-of-the-art facility at the Eastern Complex.

By the time Stewart had fully checked over the valves and fittings, verified cylinder pressures and run a quick analysis on their contents, *Skimmer Two* had settled into the warm waters of the artificial harbor at Millersville, centered on the western shore of Jarvis.

A century ago Millersville was a "thriving" community of five or six people put in place by President Roosevelt to establish residency on the island, but with the start of World War II and shelling by a Japanese submarine, the island was abandoned. Eighty years before that, Jarvis was a going guano mine first for Americans, then Kiwis from New Zealand, and finally British. On August 13, 1913, the barquentine *Amaranth* ran hard aground and broke up on the south shore with a load of coal bound for San Francisco. The crew and passengers eventually found their way back to civilization, but even now lumps of black coal occasionally wash up on the south beach.

The occasional Jarvis residents for the last century and a half would never recognize what Jarvis was now. The runway along the south shore with its western apron and sparkling white hangar buildings defined the southern end of the island. The eight-meter-high berm along the northern and western shores was still there, broken at Millersville and again near the southern end where the old rail lines had penetrated to the beach for transferring guano to waiting schooners. Instead of rail lines and the old Millersville small boat landing, a deep-water wharf paralleled the western shore right at the edge of the coral reef, connected to the shore by a wide, concrete-surfaced elevated roadway. It could handle all but the very largest ships afloat. The floating barriers forming the artificial harbor boundaries protected the wharf. The white buildings of the recompression complex occupied the stretch between the southern cut and the Millersville wharf, all connected by a concrete two-lane road. Atop the berm along the north shore, the main Jarvis Compound sparkled in the early afternoon sun. And at the island center near the old guano diggings lay the foundations for what would eventually become the socket for *Jarvis Skytower*—downlink of the doorway to the stars.

Carver eased *Skimmer Two* next to the waiting floating dock connected to the wharf. With line in hand, Gofort stepped from the stern well onto the dock and secured the stern to a black cleat several yards behind the skimmer. While Carver kept the bow hard against the dock, Gofort stepped across to the narrow walkway ahead of the cockpit, retrieved a line from a flush locker, and secured it to another black cleat several meters ahead of the gently rocking bow. Carver cut the engines, and as the whine faded, the only sound left was the rolling surf outside the breakwater, and the thousands of birds that still made Jarvis their home.

Carver flipped a switch, and the entire starboard side of the skimmer opened like a clamshell. A white pickup drove up the wharf with Stewart at the wheel.

"Let's get those KPs offloaded and up to the chamber," he said to Gofort as he waved to Carver. "Hey, Jeff!"

Fifteen minutes later, Gofort was sitting inside the Jarvis chamber facing a livid Watson and his two companions, Tiffany and Carey. All three were breathing pure oxygen through small masks that cupped their noses and mouths, while Gofort breathed the chamber air.

"You people could have killed us!" the angry Watson sputtered through his muffling mask while Tiffany and Carey sat quietly listening.

"Ye waur sabotagin' th' project," Gofort answered back, his Scottish lilt punctuating his words. "We could hae jist towed th' *Green Avenger* back tae Jarvis withit ye," Gofort said. "Hoo lang dae ye think ye would hae lasted?"

The two young people looked at him with wide eyes. "You wouldn't have done that," Carey sneered. "There're laws against killing people."

"Whit dae ye mean?" Gofort countered. "What laws?"

"U.S. law…Jarvis is part of America…" Carey's voice was high-pitched, almost squeaky as his tension mounted.

"Aye, Laddy, but we waur nae oan Jarvis, waur we noo?" Gofort paused, glancing at each of his patients. "Th' reason y're living at aw is th' Boss, Alex Regent. Ah would hae left ye an' chummed th' water oan mah way it."

Tiffany's eyes got even wider.

"That's reit, Missy…shark bait, that's whit yoo're guid fur."

"You're the abomination," Watson quipped.

"Whit dae ye mean, abomination? *Slingshot* is a green project, fer chrise sake. Nae pollution, nae rockit exhaoost, nae e'en nukes. They don't gie mair green than *Slingshot*."

"We destroyed this planet, so we go to the next?" Watson said with a sneer. "Look at your damn skimmer. My *Green Avenger* runs on wind." Tiffany and Carter nodded.

"But we're gonnae to make it possible tae move pollutin' manufacturin' aff Earth—tae th' Moon ur th' asteroids, even…" Gofort had his own passion.

"So you can pollute them too," Watson interrupted. "It's not enough you did the greenhouse thing here, now you got to do it out there too?"

"Greenhoose…whit dae ye mean? We're oan a solar-driven coolin' cycle reit noo; back then was th' oopslope ay th' cycle."

"That's just multi-national big company propaganda," Carry interjected. "WE stopped global warming, and THEY give the Sun credit." Gofort could hear the uppercase letters.

"It's all for profit, you know," Tiffany said, "and we pay the penalty."

"A million sooty tern pairs made Jarvis their home," Watson said, "and fourteen thousand boobies…thousands of lesser frigatebirds…" His voice trailed off. "What about the eastern coral reef?" he asked. "You're destroying all of it…and for what? So you can go and destroy another world?"

Watson turned to look at his companions, and his eyes crinkled in an apparent smile. He started to remove his oxygen mask. "Leae it in place, Sairrr, ur I'll tape it tae yer face an' tie yer hans behin' ye, is 'at clear?"

"You can't do this to us!" Watson insisted. "I'll sue you!"

"Hoo much mair of thes shit, Stu?" Gofort asked over his headset.

"Fifteen minutes, Gofer."

"Whit dae we dae wi' 'em then? We got nae lock-up."

"Wait one…"

Stewart contacted Alex and posed the question. "Keep them in the chamber at three meters," Alex said. "Give them a piss bucket, and feed them through the medical lock 'til I get there tomorrow. I don't want them anywhere on the island."

"Roger that!"

Stewart explained the plan to Gofort. "You and Jeff on chamber watch, port and starboard," he said. "Tex and me are headed to the Eastern Complex. We're already late."

The chamber surfaced, and Carver pushed the door inward and handed Gofort a bucket. Gofort turned and announced that the three occupants would have to remain inside the chamber for the night, for their own protection.

"But I have to pee," Tiffany said.

"Use th' buckit, Missy," Gofort said as he left and closed the heavy door behind him. Then Stewart pressurized the chamber to three meters, making the door impossible to open from the inside.

While the other three were dealing with their charges, Sanchez refueled *Floater Two* and grabbed a bite to eat. He remembered to make enough so that when Carver showed up for food, there actually was some available.

"Hey Tex!" It was Stewart on the SSRS. "How soon can we leave? I got to get out to the complex."

"Whenever. I'm fueled and ready to go. Soon as you get here." He lay back in the shadow under *Floater Two*, for a short nap. It had been a long day, and it wasn't over yet.

A half hour later Stewart showed up, and ten minutes after that they were full throttle on a due east course for the Eastern Complex.

"Eastern Complex—it's Tex and Stu. We're in the air, full bore. How're things?"

Down on the complex, the SSRS signal automatically routed itself to Cody Haydon, Lead Engineer for the Eastern Complex. "Got the press on hold. Your guys are standing by. Alex, Klaus, and Margo are in the air…be here in a bit. What's your ETA?"

"'Bout twenty minutes or so, Cody."

"Okay, Tex…keep an eye out for Alex. Don't want a collision causing an accidental fireworks display for the press."

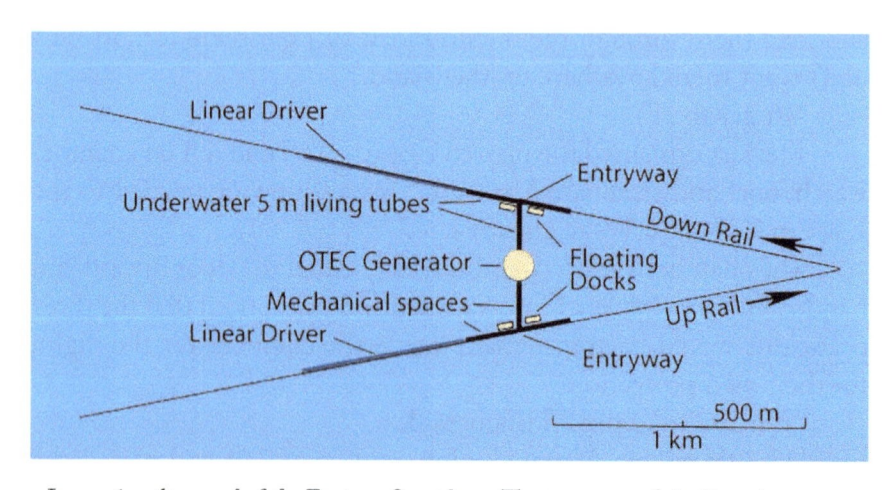

Image 6—Apex end of the Eastern Complex—The two arms of the Complex meet at the far side of a twenty-kilometer-wide circle about fifty kilometers further east

EASTERN COMPLEX 300 KM EAST OF JARVIS ISLAND

Three-hundred kilometers to the east, Alex banked *Floater One* into a sharp turn over the apex of the Eastern Complex. Looking down he could see the encasing tube floating on the surface, ready to be submerged. Two arms stretched from the apex to the east to meet at the far side of a twenty-kilometer-wide circle about fifty kilometers further east. The huge circular mouth of the OTEC generator nestled inside the apex, forming a windbreak so that the surface within the apex to the west of the OTEC cylinder was completely calm. Alex spiraled in, setting *Floater One* gently on the glassy surface, and taxied to a floating dock tight against the inner side of the tube.

Haydon was waiting on the dock, with a female reporter and a cameraman sporting one of the latest in portable holocams with miniature pickups mounted on each shoulder and a transparent holographic viewer positioned before his right eye. Haydon bent a line to the nearest pontoon. His serious face cracked with a brief grin that faded nearly as quickly as it appeared. He waved and then brushed a hand over his short-cropped brown hair, and repositioned his aviator-type sunglasses. He had a boyish look for all of his forty-five years—perhaps resulting from his 1950s-style flattop.

Under the watchful eye of the cameraman, Haydon offered Margo a welcoming handshake, nodding briefly as he spoke her name, "Margo."

"Cody."

Klaus opened the rear door, jumped to the dock and boomed, "Cody—sorry we're late, but I'm sure Stu or Tex briefed you already." They shook hands heartily while Haydon nodded in the affirmative.

Alex slid across to the dockside door and jumped to the dock. "Alex," Haydon nodded again, shaking his hand. "We've been keeping busy while we waited, but," turning to the reporter, "the news crew is getting a bit antsy." He gestured in the general direction of the female reporter and her cameraman against the backdrop of the OTEC well and the floating tubes beyond.

"Loraine Kutcher, please meet Alex Regent, the Boss, Margo Jackson, underwater construction, and Klaus Blumenfeld, power systems."

The reporter stepped forward, shaking each hand in turn. "I'm Lori, and this," turning to her cameraman, "is Dex Lao, the best cameraman anywhere."

Lao simply nodded, stepping back to take in the scene with his holocam.

Alex acknowledged the introduction, and faced toward the OTEC, arms spread wide. "I presume," he said, "that Cody has given you a once over." It was a statement, not a question.

❋

Lori turned toward the water as Lao smoothly swung his shoulders. She was short but well proportioned with shoulder length blond hair. She wore a short blue skirt with a white open-neck blouse that exposed the tops of her surgically enhanced breasts. Her eyes were startlingly blue, and her ambition was boundless. Hailing from Seattle, she had worked her way up the Fox Syndicate ladder, using every asset she had at every possible turn. She was beautiful, smart, and ambitious, and this was her first big break. She had drawn the long straw (at least that was how she related the story) as the *Slingshot* pool reporter for the coming month, and she intended to make the most of it. Already she knew that Alex was deliberately keeping himself formal (but she intended to change that), and Margo had actually caused her stomach to flutter—another avenue she intended to pursue. Klaus and Haydon she decided to keep in reserve.

Lori's eyes first were drawn to the sides of the OTEC well that extended some four meters above the surface. Her eyes and the holocam then followed the floating walkway against which they were moored. It marked the underwater location of the northern five-meter-wide tube that housed the living quarters, below which was suspended the vacuum-sheath-encased ribbon surrounded by the induction motor. The walkway passed to the right of the vertical OTEC well, with a branch leading right up to the well wall, and continued for a hundred meters to the end of the living quarters. There the tube crimped down to a rigid three-meter-wide tube that housed only the ribbon, the down-slope linear induction motor, and an access tramway. It angled to the surface and ran for a full two kilometers more, blending with the ocean surface despite its fluorescent-orange coating. Although they could not see it, from there the tube, containing only the vacuum sheath with the ribbon and the steering magnets, continued for forty-eight kilometers. Then it commenced a slow thirty-two-kilometer arc to the north, forming the outer rim of a twenty-kilometer-wide circle, meeting with its identical counterpart from the northern up-slope segment. The northern tube mirrored the southern one, passing behind OTEC from their vantage point. It only differed from the southern extension in that the up-slope linear induction motor drove the ribbon toward instead of away from the apex, and there were no living quarters.

Overhead, *Floater Two* soared out of the afternoon sun, engines buzzing.

"Hey, down there," Sanchez said over the SSRS, rocking his wings.

All eyes turned to the sky as Sanchez tipped *Floater Two* into a steep dive, and deftly set her down in the clear area west of the OTEC. He taxied to the dock ahead of *Floater One*, where Klaus and Haydon set the mooring lines.

Lori watched Sanchez and Stewart debark from the plane and join the group on the dock. Alex made introductions, and Lori allowed her hand to linger with each of them as long as possible without being obvious. Then Stewart hurried off to supervise his divers during the submergence. Sanchez stayed with the group.

※

Alex turned to the reporter and then to Haydon and smiled. "Lori, Cody…shall we?" He gestured to the others to fall in.

This submerging operation really fell into the combined purview of both Klaus and Margo, since Klaus had charge of the overall OTEC and linear driver systems, and Margo ran underwater construction. Margo had cut her teeth early on in the project supervising the assembling and submerging of 1,828 kilometers of double rail that stretched between Baker and Jarvis Islands. After that, of course, she built the one-hundred-thirty-four or so kilometers of tubing that comprised the Western Complex, and worked closely with Klaus during the construction of the two western linear drivers and the Western OTEC.

Margo had worked closely with Haydon on the Eastern Complex, ensuring that the lessons learned during construction of the double ribbon and the Western Complex would be applied here. Alex was fully confident that the next few hours would be without significant incident. He let his eyes drift toward the petite reporter and then to Margo. He sighed. *That's the last thing I need right now!*

"Let's take a few minutes," Alex said to everyone while looking at Haydon, "to review the present status of the entire project, and then explain the operation you will be observing this evening." He smiled at Lori. "Let's bring you two up to speed."

Haydon led them through a lock that was identical with the entry lock on the Western Complex, down into the pressurized living quarters, and through an internal lock into a relatively spacious conference room. At nearly three meters high and eight meters long, it was easily the largest open space in the Eastern Complex. It occupied the entire width of the tube, and both sidewalls consisted of Lexan with a refractive index the same as seawater, so that the walls were virtually invisible.

Lao stopped at the entrance and slowly turned to record the entire scene. Then he quickly stepped back into a corner to record everyone entering the room.

A long table occupied the center of the room, with sufficient chairs for everyone. Haydon tapped his wrist Link, and a shimmering image took shape over the center of the table, coalescing into a three-dimensional view of the Eastern Complex.

Haydon cleared his voice. "Record what you wish," he said to Lao. "I'll get you a holofeed of this stuff." He indicated the floating image.

Everyone sat down, except Lori who moved alongside Haydon, while speaking to her worldwide audience. "I'm speaking with Cody Haydon who is the lead engineer in charge of the vast floating complex we just saw before coming down here. Cody, please explain to my viewers what we are about to see." She smiled encouragingly at him.

Haydon tapped his Link to generate a bright red pointer that he used to indicate the elements he was discussing.

"Why don't we start with an overview of *Slingshot* itself," he suggested.

Another tap—the image shimmered again, and the view zoomed out to display Baker and Howland Islands at one end and Jarvis at the other. Stretched between Baker and Jarvis was a line that looked like a string of pearls. Beyond the islands, teardrop-shaped complexes outlined with pearls extended to the west and east respectively, pointing at the islands. "This is *Slingshot* in its current state," Haydon said, moving his pointer around the ribbon path. "A continuous tube of segmented soft iron suspended inside an impervious vacuum sheath extends for eighteen hundred twenty-eight kilometers from Jarvis to Baker, around the Western Complex," his pointer traced the complex a second time, "and then back to Jarvis and around the Eastern complex. Fixed and electro magnets bend the tube around here," Haydon swept the pointer around the Eastern Complex, "along what will become the upslope," he pointed to the apex, "back to level," the pointer flashed over Jarvis, "along what will become the down slope," the pointer flashed over Baker," back to level," the pointer indicated the western apex, "around here, back to horizontal, up and over, and back down to the Eastern Complex."

"Incredible, absolutely incredible," Lori said, beaming at Haydon. "How big is this iron tube?"

Haydon tapped his Link, and the overall view dimmed, and was superimposed by a detailed exploded view of the tube in its casing. "About five centimeters across, as you can see here," he said, moving the red pointer to the exploded view. "See how the short segments

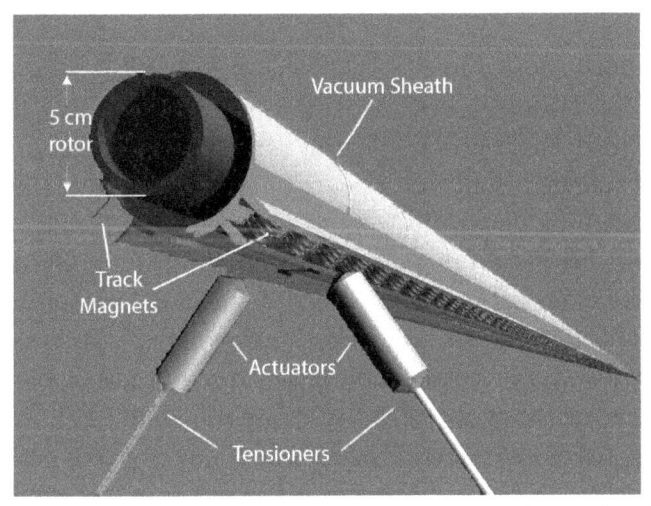

Image 7—Close-up cutaway of the tube in its vacuum sheath at 80 km altitude

overlap each other male-female fashion, sliding in and out to adjust the overall length? This whole thing is about the size of a silver dollar."

Haydon tapped his Link again, and the exploded wire view disappeared. The complex legs leading into each apex flashed. "The ribbon is driven by four linear induction drivers here, here, here, and here." He indicated the flashing sections. "When we have finished this presentation, we will lower the casing here at the Eastern Complex to its permanent position at thirty meters. We have already anchored the casing to the bottom with anchors spaced about a hundred meters apart." He looked around the table.

Lao gestured for Haydon to look directly at the holocams. Haydon turned and smiled at Lao. "That's five hundred anchors," he said.

The scene blinked, shimmered, and coalesced into a bottom view, a close-up of one of the anchors. It was a five-meter cubical concrete block anchored to the bottom with three concrete pylons. Another shimmer, and the view shifted to a cross section image of the anchor block and its pylons, extending ten meters into the bottom. Each pylon was a meter thick with a pointed bottom, and each had three two-meter long bronze paddle-like braces jutting upward at forty-five degrees, designed to prevent the anchor from being pulled out of the bottom.

"It's not very likely to move," Haydon said. "Not many storms down here, but this will weather even a big one."

"How long did it take to install the anchors?" Lori asked.

"Over a month," Haydon said.

"Oh my…"

The view shifted to the 1,828-kilometer double ribbon. "This," said Haydon, "is the part that will lift to a height of eighty kilometers above sea level. Right now it's floating at thirty meters below the ocean surface, suspended from this series of buoys a bit south of the islands." He indicated the length from Jarvis to Baker. Then, as he spoke, the encased double ribbon rose up to its extended height. "We drop elevator cables from here and here, called skytowers," he indicated points immediately above Jarvis and Baker, "and anchor them to sockets on the two islands." Anchor cables dropped from the indicated points, and a physical structure appeared at the intersection point over the islands. "These are the *Amelia Mary Earhart* and *Frederick Joseph Noonan Skyports*—*Earhart Skyport* and *Noonan Skyport* for short. And the skytowers are called, of course, the *Baker* and *Jarvis Skytowers*."

Lori looked at Alex with her big blue eyes. "Alex, why do you have to remove the protective casing around the double rail, and how do you do this?"

"A good question, Lori," Alex said. "The soft iron tube is suspended inside the vacuum sheath. Together they make up the rail. Launch capsules need direct access to the rail in order to couple to it magnetically. Underwater and in the atmosphere, the rail is encased in a semi-rigid lightweight casing made of Kevlar and other stuff. It is kept pressurized with nitrogen by separator pumps located at intervals along the casing, and the casing is compartmentalized so that if one compartment floods, the rest will remain dry. That permanent casing extends from each of the skyports down and around each complex. Before we commence operation, the nitrogen will be evacuated from the casing so that the soft iron tube is doubly protected—by the vacuum sheath and by the evacuated casing. Since the environment in space between the skyports is a virtual vacuum, we can forego the protective casing between the skyports so the capsules can magnetically couple to the rail for launch. This casing consists of a special polymer we developed that is strong, impervious to air and moisture, but when saturated with ozone and then exposed to ultraviolet, disintegrates to the molecular level." Alex smiled at Lori, who appeared to be listening with an intensity approaching rapture. He glanced at Margo briefly, but could not tell if she was amused or upset by Lori's intensity. "As the rising casement passes through the ozone layer, rarefied though it is, sufficient ozone penetrates the polymer so that when it reaches the top of the atmosphere above the ozone layer, the ever-present ultraviolet causes it to disintegrate in a matter of minutes."

Lori started to ask a question. Anticipating it, Alex continued, "Of course we have to schedule the rise so that it reaches the ultraviolet level during daylight before the ozone has a chance to bleed out of the casing." Alex nodded at Haydon.

"Any questions?" Haydon asked.

"Alex, how long does it take to erect the loop?" Lori asked, pulling her shoulders back to give the best possible exposure to the tops of her breasts.

Haydon reversed the time line so that the double rail in its casing was back underwater, while Alex winced inwardly, maintaining his focus, not letting his eyes leave Lori's face.

"The magnet deflectors at the apexes start out near the islands," Alex said. "As the loop comes up to speed, we start pulling them back and increasing the angle. This causes the main casement between the islands to begin rising." While Alex explained, Haydon manipulated the display. The double ribbon began to rise. "As the casement rises, we attach tensioners that are anchored to the bottom just like the anchors for the Eastern and Western Complexes." Cables angling out and down to the left and right from the rising casement appeared. Alex smiled at Lori, keeping his eyes focused on her face. "It takes about ten days for the entire loop to reach full speed and proper altitude."

"So, if there are no further questions, we'll get the submersion process underway."

He nodded to Margo who stood up.

"Let's go do it!" Margo said.

Everyone filed out and back through the lock to the top of the living quarters tube.

"You and Margo supervise from down here," Alex said to Klaus. "I'll take our guests up to get some good news shots from above."

In a few minutes, *Floater One* glided away from the tube and shortly was airborne. Lao occupied the front seat with Alex so he could get good holoshots. Lori had to content herself with a back seat, but that gave her the opportunity to lean over Alex's shoulder from time to time to ask questions.

JARVIS ISLAND RECOMPRESSION COMPLEX

Three hundred thirty kilometers west, Gofort smiled at Carver and rose to his feet. "Three meters an' lots ay bitchin'," he said as he officially turned over the watch to Carver. "I'll roon up some sandwiches an' somethin' tae drink."

"Don bring me nothin'. I already had sumpin', but I left you a stack o' samiches." Carver turned his attention to the dive control console. He punched up the communicator.

"In the chamber, hey—we be sending in some samiches in a bit. You guys want sumpin' to read?"

"Get us out of here!" Lars shouted. "We got rights."

"I got to pee again," Tiffany wailed.

"And I got to take a dump," said Carey.

"Use the bucket, Girl," Carver told her. "You, too, Carey."

"But the guys," Tiffany wailed again.

"Deal with it," Carver told them. "Work it out. You still got 'nother twelve or mo' hours in dere, so get used to it."

CHAPTER NINE

AIRBORNE ABOVE THE EASTERN COMPLEX

Alex and his charges circled at about 2,500 meters. Dex recorded continuously.

"What you're gonna see," Alex said, "is that the entire floating tube extending out from the living quarters on both sides and around the loop will settle underwater." He turned and smiled at Lori. "Sorry, no drama."

"Alex to Klaus; what's the status?"

"Margo is setting up the tensioners right now. We should be ready to go in about five minutes."

"Remember the anchors we showed you?" Alex asked Lori. "They're attached to anchor points on the underside of the tube with the tensioners."

Lori nodded and squeezed Alex's shoulder.

"Look!" Alex pointed.

The entire Eastern Complex was slowly sinking beneath the gentle waves.

"You getting this, Dex?" Lori was nearly breathless.

As they circled above, watching, the fifty kilometers of tubing took on a shimmering look as it slipped beneath the water, and then it was gone from sight.

"Ten meters," Klaus announced over the SSRS.

"Twenty meters."

"At depth."

"That may have seemed routine to you," Lori said to Alex, "but it was incredibly exciting to me." She ran her hand through Alex's hair. Lao pointedly focused his gaze out the window.

Alex found himself at a loss. They didn't get more beautiful than this sexy female on the make, and he was frustrated following weeks on the project, pining away for cool-as-a-cucumber Margo. But he also knew Lori's game. She wasn't interested in Alex—only the "guy in charge," who happened to be Alex. He also was pretty sure that her intense interest was only one more way to advance her career. Nevertheless, he resolved not to turn it off. Perhaps he would even take advantage of her attractive offer before she departed. While Alex was not prone to notch his pistol grip, he was nevertheless willing to enjoy an occasional sideshow—or in this case, perhaps, a short-lived main event. He pulled himself back to the current show.

"How are the tensioner tests coming?" he asked Margo.

"Completing them now," she said. "We found a couple that need attending, but we expected that. Stu's guys are on their way to correct them now." After a pause she continued, "What's going on up there?"

Alex felt soft fingers trailing across the back of his neck and down the front of his shirt. "A bit of sightseeing," he answered. "We'll be back soon."

Lao continued to focus his gaze out the window.

※

Two hours later, as evening approached, Alex feathered back his props and rolled to a stop on the Jarvis tarmac. He had left Lori with Margo and Klaus, not wanting to expose the operation's troubles with *Green Force* to the media—at least not yet.

Carver met him as he shut down *Floater One*. "You got a whole lot o' bitchin' in the chamber," Carver told Alex, offering him a wide grin. "Dat Lars laddie, he pissed, fo' sure."

They climbed into the white pickup for the short drive to the chamber complex.

"How di't go?" Carver wanted to know.

"Couple of minor problems. We're on top of it."

Just then: "Alex, it's Margo."

"Go, Margo. What's up?"

"We've got problems, Alex. Big ones."

"I'm listening, Girl…"

"The first tensioner problem was just a flooded power connector. We fixed that in a half hour." She paused. "But the other problem…" She paused again. "The other problem is that someone cut the tensioner cable. Not all the way through, but it's seriously weakened. The first big storm, and it'll separate."

Alex sat quietly digesting the news.

"Stu installed a double clamp splice, but it won't hold forever. Alex, we've got to replace the cable. That means either another anchor or a minisub dive to fix this one."

"Okay, Margo. You and Klaus work out the best approach and get back to me." He looked out the pickup window as Carver pulled up to the chamber complex. "I'm going to be busy for a while with Lars and his crew." And then, "Think he did it, Margo?"

"Who else?"

JARVIS ISLAND RECOMPRESSION COMPLEX

Alex strode into the chamber room followed by Carver.

"Gofer." He nodded. "Bring 'em up."

Gofort gave Alex a thumbs-up, and turned to the dive console. "Surfacing," he announced to the chamber. A moment later a loud rush filled the room as the chamber internal pressure equalized with the equatorial sea surface pressure. Carver cracked the round hatch, and Watson stumbled out followed by the other two. Alex stood there looking at them with obvious disgust.

"Who the fuck are you?" Watson ventured before Alex stopped him with a raised hand. The human smell emanating from the chamber and the three Greens was overpowering.

"Take them to the showers and let them clean up," Alex told Gofort.

"But we don't have anything clean to put on," protested the young, petite girl sporting long brown hair that was decidedly a mess. She spoke in a clear voice with good diction.

"And you are?" Alex asked.

"Tiffany…Tiffany Montague, and my daddy's going to have a piece of you when this is all over."

"I'm Alex Regent, Tiffany, and I run this show, all of it. That puts me one small step down from God." Alex smiled at the pretty girl and asked, "How old are you?"

"Eighteen—Lars, do something!" She turned to the lanky, nearly ascetic Watson, who just shook his tousled yellow pageboy helplessly.

"Get your pretty eighteen-year-old butt into the shower," Alex said, "and wash that stink off you and your clothing."

"Wear wet clothing?" Tiffany protested.

"You'll dry soon enough," Alex told her. "Now get!"

Alex turned to Carmine Endsley. "Who're you?"

"Carey…and fuck you and your asshole project! Fuck you…fuck you!" The second expletive came out with a squeak. His dirty long hair was tied back in a ponytail, and he sported a spotty full beard. He reminded Alex of photos he had seen of 1960s hippies.

"Go clean yourself up," Alex told him.

"Fuck you," Carey said again with an emphatic squeak.

"You know what a G.I. shower party is?" Alex asked.

Carey just glared at him.

"Get your scrawny butt in there right now, or my boys here will carry you in and scrub you raw with a floor brush." Alex didn't raise his voice.

Carey capitulated and headed to the shower, grumbling.

"And you, Mr. Watson…you're too smart to turn down a shower." Alex looked him over. "Am I right?"

Watson turned and followed Carey into the shower room.

While the motley crew cleaned up, Alex briefed Gofort and Carver.

"An' whit dae we dae wi' these idiots?" Gofort asked. "An' th' others oan th' schooner?"

"They're mostly kids," Alex said, "enraptured by the charm and wit of one Lars Watson." Alex thought for a moment, and then made up his mind. "We're going to hold the kids for retrieval by their parents." He grinned at them. "We'll deal with Mr. Watson later. And I have some thoughts about the Endsley guy—the computer whiz."

Tiffany appeared first, actually looking fairly presentable, her long brown hair hanging straight past her shoulders, framing her naturally pretty face.

"You're looking and smelling better," Alex said to her with a warm smile.

"Thanks…I guess." She looked down at the floor, devoid of her initial fiery anger. "What happens now?"

"Well, I contact Mr. Montague, and Mr. Montague makes a personal journey out here to pick you up." Alex gave her a warm smile. "How does that sound?"

"You've got to be kidding. Daddy would NEVER do that." Alex could hear the capital letters clearly. "He'll send someone—he always does."

"Not this time," Alex said, "not if he wants to see his little girl anytime soon."

Tiffany started crying quietly.

Carey strode out, dripping wet and barefoot. "Okay, I washed. Now what?" he squeaked.

"Your family's in Palo Alto, right?" Alex asked mildly.

"So what."

"You hack into my system—fuck with some drawings?" Alex kept his voice even and calm.

"So what if I did?"

"Pretty good hacking," Alex said. "You don't know shit about engineering, but the hack job was impressive."

Carey straightened a bit and looked up at Alex. "Really? You think so?"

"'Sa fact," Alex said. "Really."

Watson strode in, face full of fury, apparently having heard Alex's exchange with Carey.

"Remove him down the hall and lock him up," Alex told Gofort. "Make sure he's secure."

Alex turned back to Carey as Gofort pushed Watson through the door. "I could use a guy with your skills."

"Really?"

"What does Watson pay you?"

"Are you kidding? Nothing. I have a family trust." Carey's voice adjusted to a lower register as he calmed down. "I'm helping Lars save the planet; that's what it's all about—saving the planet."

"Saving the planet… Really?" Alex was actually beginning to enjoy this. "And how's that?"

"We save old-growth forests…we save rain forests…we save coral reefs…we save endangered birds…polar bears…fresh water fish…" His voice commenced squeaking again.

"I see," Alex said, "so tell me how you save old-growth forests."

"We drive steel spikes into the old trees so they won't be cut down."

"Really…how does that work?"

"Well sure…if a logger cuts down a spiked tree, the spike can break his chain saw so he gets hurt. So instead, loggers don't cut down old timber. It's simple."

"Is it? Do you tell them which trees you spiked?"

"No."

"So, you're okay with some guy losing an arm or leg or his eyes, so his family can't survive? You're okay with that."

"That's not what I meant."

"I see."

"Really, nobody wants that—we just want to save the trees…" His voice trailed off.

"I see," Alex said. "And what about the rain forest—the jungle, because that's what it really is? Which one?"

"Amazon."

"Really?"

"Yep."

"How big is this rain forest—this jungle?"

"Big…really big. Big as Texas."

"I see. And the reason?"

"It's a habitat." Carey's voice was squeaking again.

"For what?"

"Critters…things…plants…it's important!" Carey appeared to think. "It's for the O_2., that's it…for the O_2."

"Did you know that most of our O_2 comes from plankton in the ocean—not the rain forests?"

"No way!"

"Way."

"Really?" Carey sounded genuinely interested.

"'Sa fact. Much of the South American jungle is a disease-ridden, festering swamp. Some of it we probably want to retain, as a continuing habitat for the critters that live there, and also for what we can learn and benefit from. But the rest—if it's cleared, it's land that can feed a lot of people."

"Really?"

"Yep." Alex smiled at him. "What about the coral reefs?"

"We're saving them too. Got to. Humans are killing them."

"Really? Who says?"

"Well, Lars says…"

"Right…he's a real authority, isn't he?" Alex softened his voice a bit. "Coral reefs…they come and go, you know. It's just part of Nature's cycle."

"Really?"

"Yep. Not only that, whenever people like you try to interfere with a dying reef, they don't change anything, and sometimes they speed up the death."

"No way!"

"Way." Alex paused. "But the reefs always come back…always have, always will. It's way beyond your or my paygrade."

"No shit…"

"What about the birds, polar bears, fresh water fish…tell me about these."

"Well, what about the birds right here on Jarvis? They're endangered by this operation."

"You mean the two million sooty terns? The fourteen thousand boobies, the thousands of lesser frigatebirds? Are those the ones you mean? Cause there are more now than ever before."

"No shit…"

"And polar bears are thriving, as many as there have ever been. And the fresh water fish…well perhaps we need to be a bit careful there, since there is an absolute limit to how many can breed. So maybe we can do something about them, and that's one out of how many?" Alex looked him directly in his eyes. "And for all that Lars is paying you nothing? And you are not accomplishing anything useful to anybody or anything? And now you're in deep doodoo?" Alex stepped back. "Are you kidding me? Are you really that dumb?"

Carey stood there silently. Alex walked up to him and took him by his shoulders. "What if I paid you a lot of money to do some really interesting computer stuff. Stuff that's never been done before. Stuff that'll make you world famous." Alex stepped back again. "What if I gave you a job where you can help change the world, make it a better place, a cleaner, greener place?" Alex stepped into his presentation mode.

"We're going to move polluting enterprises off-planet, to the Moon and asteroids. We're going to open up the solar system to ordinary men and women, so regular Joes can go anywhere, do anything, be what they want to be. This is what's happening here, Carey, and I am offering you a part of it." Alex paused and lowered his voice. "What do ya say?"

Carey grinned, and squeaked, "Where do I sign up?"

※

Tiffany stood quietly sobbing. Alex began to feel sorry for her. He walked over and put his arm around her. "Don't you worry, kid. We'll get you home soon enough."

"But I want to stay and help," she said, drying her eyes.

"You have nothing to offer," Alex told her seriously, "nothing." He squeezed her. "Go home, finish your education, figure out what really matters, and get on with your life." He smiled at her. "If you really like what you heard here, learn something useful and become part of it. You're young, and this is just the bare beginning." Alex grinned and looked around the room. "It's open ended from here. There are no limits." Alex turned to Gofort. "Find them some bedding and let them have a good night's sleep. We'll take care of the rest in the morning when the schooner arrives."

He strode out toward the room holding Watson.

On his way, he Linked through to *LLI* Chairman Mabel Fitzwinters. When she answered, he said, "Mabel, it's Alex."

As soon as she recognized Alex, Mabel came fully alert. "Go ahead, Alex. What is so important this time of night?"

Alex briefed her on the situation, and informed her about the capture of apparently the entire gang. He briefed her on his intentions for the kids, for computer whiz Carmine Endsley, and for Watson.

"There's more to this than you may realize, Alex," Mabel said. "We just identified the possible source of your local troubles out there." She paused, glancing down at something off camera. "As soon as I know more, you'll be the first to know. In the meantime," she added, "I've got your back." Alex disconnected.

※

"Lars Watson," Alex said as he strode into the room where Watson was confined. Watson was sitting in a straight-backed chair at a wooden table. He looked up at Alex defiantly.

"Why, Watson? Give me a reason—one that makes sense," Alex said as he sat down opposite Watson, keeping his voice conversational.

"Cause you're destroying the planet, Man."

"What do you mean, destroying the planet? This is the greenest large engineering project ever."

"The biggest con job, you mean," Watson interrupted, passionately. "You're destroying millions of birds on Baker and Jarvis, sea turtles, the coral reefs on both islands; you're even decimating the local shark population. What's green about that?" Watson pounded the table with his fists. "What's green about that?"

"Watson, you're either ignorant or stupid. In either case, you don't begin to understand what we are doing here." Alex let a bit of frustration creep into his voice. "There are more birds on Baker and Jarvis than there ever were, precisely because we're here. Our two OTEC systems bring up water from a kilometer down. It's nutrient-laden, and the fish love it. Consequently, lots more fish—bird food." Alex shook his head in amazement.

"We don't touch the coral, except for our boat landings and docks, and even there, over time we enhance the coral base. Our structures float, except for the anchors which sit in the silt five klicks down." Alex folded his hands on the table and looked at Watson earnestly. "Virtually everything we do enhances the local environment. Things here are a lot better for virtually every species since we commenced *Slingshot*."

"Yeah—what about Howland?" Watson's voice took on a sneer.

"Howland—we really fucked that one up, Lars." Alex's voice dripped with sarcasm. "Back in 1937 the U.S. Government built three runways on Howland: sixteen hundred meters running north-south, seven hundred thirty meters east-west, and nine hundred fifteen meters northeast-southwest—back in 1937. The Japanese bombed and shelled the hell out of it in late 1941. From the 60s, everyone pretty much left it alone, except for occasional visits." Alex stood and walked across the room.

"So, *LLI* rebuilt the long runway to handle large planes, and the other two for increased use. We added roads and deep-water moorage, which means we sunk pylons into the edge of the reef." Alex turned and looked at Watson. "But so what? We did what humans have always done. We modified what nature gave us to better use it for human benefit. In the process, we destroyed a few things, but we created more than we destroyed. Already, after only a couple of years, the coral around the Howland deep-water docks has recouped, and is growing faster than ever. The birds never did leave, and we have to scare them off with ultra-sonic acoustics whenever a plane lands or takes off." Alex sat back down at the table facing Watson.

"So enough of your ridiculous complaints!"

Alex leaned his chair back remembering the congressional hearings before the opening of the national refuges of Howland, Baker, and Jarvis Islands. The zoo from three years earlier had occupied the better part of three months of his life. He and Mabel Fitzwinters, the tough chairman of *LLI*, must have testified before a dozen committees and subcommittees. Despite the best efforts of board member Rex Johnson, who had forgotten more about congress than most members of Congress knew, they seemed to be up against a brick wall until they brought Margo into the equation. Margo had charmed the jaded old geezers into submission. After that, Congress quickly passed legislation that granted full economic exploitation of Howland, Baker, and Jarvis Islands. The governing rules were simple enough: keep it as pristine as possible, but don't delay construction needlessly.

The subsequent court fights were spectacular. Somehow, board member John Boyles managed to get the matter on the Supreme Court docket within a few short weeks of the passing of the legislation. Even though the Supremes were still dominated by a liberal majority established during the Obama years, their decision was unanimous. Common sense won out over senseless principle, and the project commenced in earnest.

"You cost me several hundred thousand dollars," Alex told Watson, pulling himself back to the room. "We may have to dive down to the anchor you sabotaged to replace a tensioner."

Watson looked at him sullenly. "Good. I hope your sub sinks."

Alex barely controlled his reaction; Margo would be piloting. "Your sabotage days are over," Alex told Watson. "Since I can't patrol all eighteen hundred klicks of double ribbon, I'm going to confine you until ribbon-rise."

Watson was outraged. "You can't do that. I got my rights!" he yelled.

"Not here you don't. Not now."

"In three days, I'll have a court order to sell the *Green Avenger* to cover my costs."

Watson choked with fury.

"After ribbon-rise, you'll be dropped off—oh, I dunno, Guam, New Guinea, somewhere you can't get into trouble."

Watson stood, speechless in his anger, fists balled in frustration. "You haven't heard the last of me," he said through clenched teeth.

"Oh—I think I have," Alex said as he walked out the room.

CHAPTER TEN

EASTERN COMPLEX 300 KM EAST OF JARVIS ISLAND

Margo was frustrated and irritated by the sabotaged tensioner. She and Klaus had discussed the problem at length, and finally decided to drop another anchor. They had not yet worked out all the details of repairing a cut cable, and now was not the time. It was more efficient, both in time and logistics, to expend another anchor than to solve the repair problem and then make a dive. Besides, she had work to do out west, and didn't need the delay in getting back.

Margo, Klaus, and Lori, along with her cameraman Lao, had spent a pleasant two hours together reviewing the day's events over a drawn-out meal. The food was basic, but somehow Klaus had conjured up several bottles of a good Oregon microbrew. Alex had given her the okay to brief Lori superficially about the sabotage. She spent part of the meal outlining what had been happening over the past few weeks, culminating in the damaged tensioner today. She told Lori about capturing the culprits, but gave her no details.

It turned out that Lori was a pleasant, very bright dinner companion. Margo had been amused at the effect she had had on Alex, and noted that Klaus seemed unaffected by her obvious charms. Then again, she admitted, the reporter's charms had not been focused on him. She, herself, was glad for the distraction. It allowed her to step back, take a breath, and get her personal emotions under better control. Margo remembered the underwater "kiss" and its effect on her. She flushed slightly with the memory.

Lori, who was sitting next to Margo, reached over and squeezed her hand. "My goodness Margo, what was that all about?" she asked softly, leaning closer to her. As she leaned over, Margo had a full view of her nearly perfect breasts.

"Just a private thought," Margo said.

"I know," Lori said. "I have those too." She squeezed Margo's hand again.

Margo glanced over at Klaus, who was having an animated technical discussion with Dex, and seemed oblivious to her exchange with Lori.

"It's been a very long day, Lori." This time she squeezed Lori's hand. "Morning will arrive way too soon. May I show you your quarters, and then we can retire?" Lori squeezed back. Margo stood and said, "We're going to turn in, Klaus. See you fellows in the morning." She led Lori through the hatch and down the passageway to the two-bedroom suite they would share.

They entered the suite, and Margo showed Lori how to operate the pressure-sealed hatch.

"Would you share a glass of wine with me before we turn in?" Lori asked.

"Happy to—Let me get it. I know the layout here." Margo smiled warmly at Lori, wondering why she felt the tingling warmth in her stomach she usually felt with Alex or Klaus.

Margo handed Lori a stemmed glass half-filled with a deep red Merlot. Taking one herself, she touched glasses with Lori and said, "Two girls making their way in a man's world."

"You got that right," Lori answered.

They finished their wine and then stood. Lori walked up to the very much taller Margo, stood on her tiptoes, and then kissed Margo's cheeks—first the right and then the left. In the process, her lips brushed across Margo's. "G'night, Margo," she said softly. "I really like you."

And she stepped into her room and closed the door.

With a slightly unsteady step, Margo entered her own room, a bit unsure about what she was feeling. She undressed in the dark and stepped into the shower for a minute. While rubbing herself dry, she observed that her nipples were much more sensitive than usual. With a sigh, she lay down on the bed, pulled a cool sheet to her chin, and drifted off.

Margo dreamt that she was with Alex. He was kissing her, fondling her…it was wonderful… She drifted into slow wakefulness, to feel soft lips on her lips, on her breasts…it was sleepily delicious, and she abandoned herself to her pent-up passion…

JARVIS ISLAND COMPOUND

"Mr. Montague?" It was 6:00 AM on Jarvis, and Alex had linked up with Carleton Montague in his Park Avenue office suite in Manhattan. Montague's silver-haired regal image looked at Alex as the hologram image stabilized before him. "This is Alex Regent."

"The Alex Regent from *LLI*? "

"That's right. I'm calling you from Jarvis Island in the South Pacific."

There was a brief pause, and then Montague said, "What time is it there. Must be fairly early."

Alex smiled at the image. "I'm on Hawaii time, Mr. Montague. It's six AM." Alex paused for a moment. "Have you heard of an organization called *Green Force*, Mr. Montague?"

"Can't say that I have. Why?"

"I'm afraid you're going to know more than you ever wanted to before the day is over." Alex described *Green Force* to Montague, and detailed Tiffany's role in sabotaging Slingshot. "I'm not looking for compensation or vengeance or anything else like that," Alex continued. "Tiffany's a nice kid. She needs to be back home with you and Mrs. Montague, and in school. What she doesn't need is to be running after this ecoterrorist, getting people hurt, or—God forbid—even killed, and maybe even coming up pregnant in the process."

"And so…" Montague sounded very cautious.

Alex looked at him quietly, waiting.

"You've let her go, right?"

Alex just waited.

"You will let her go then. How soon?"

Alex remained silent.

"Well, what then?" Montague's voice exhibited some strain.

"What are you doing with my daughter?"

Alex continued to wait.

"What have you done with my little girl?"

That was the clue Alex was awaiting. "Nothing, Mr. Montague, yet…"

"What do you mean? That's my little girl you've got…" A hard edge crept into Montague's voice. "I'll send down my personal jet for her."

"No!" Alex punctuated the word.

"What do you mean? No. I'll get a court order…I'll send my security force…I'll…"

"No you won't, Mr. Montague, because by the time you can muster your resources, Tiffany will be moved to an undisclosed location under an undisclosed jurisdiction, facing international terrorism charges." Alex paused for effect again. "In this post-Jihad world, you know what that means."

Montague choked off an answer.

"On the other hand," Alex continued, "if you want to come down here yourself, by yourself, you can meet me personally. I'll show you what's going on down here, and I'll introduce you to the louse who seduced your little girl."

Montague appeared to be doing something on his desk. "I'm about nine hours away in my *Gulfstream*. Will that work for you?"

"Don't push it, Mr. Montague. I'll give you twenty-four hours." Alex disconnected his Link.

Fifteen minutes later, Mabel overrode his Link. "What in hell did you say to Montague?" she wanted to know.

Alex briefed her on his conversation with Montague. Mabel looked thoughtful, and then said, "I like that. Don't worry about any repercussions. John Boyles will handle it at this end." With a twinkle in her eye, she asked, "What do you think about letting Montague take Watson back with him?"

"I've been thinking about that," Alex said. "I think it's a great idea."

※

An hour later, Margo called to tell Alex about their decision to drop another anchor.

"Great," Alex said. "Let's get it done as soon as possible. Do you and Klaus need to be there for the drop?"

"Nope. We both need to get back."

"Okay, then. Here's what we'll do." Alex briefed Margo on the Montague matter. They arranged for Tex to fly both Margo and Klaus back to Baker, and then Tex would bring Montague to Jarvis as soon as he arrived.

"What about Lori and Dex?" Margo asked.

"Let them cool their heels until we have this Montague situation behind us. Then they can come here for a photo op with the *Green Avenger*. After that, we'll take them to the Western Complex to record the activity there."

JARVIS ISLAND WHARF

"Jonesy, what's your status?" Alex had been following the schooner's progress, but wanted to let Jones know he was not forgotten.

"'Bout an hour out, Boss. The kids are behavin', and I should have them under control when we get there. We could use a shower and some real food, and I could use a nap."

"We'll meet you at the Wharf. Watch the reef coming in."

An hour later, Alex, Carver, and Gofort waited on the Wharf as the *Green Avenger* pulled up to the floating dock. The tropical sun beat down remorselessly, unhindered by clouds. A squall that had passed earlier just to the north was already over the horizon. From where they stood, the horizon to the north, east, and south was unclouded but hazy. The ever-present easterly wind was partially blocked by the berm, so that the heat was more oppressive than it would have been on the other side.

Jones supervised as a motley crew of two girls and two boys slid the gangway across the narrow gap to the dock. They seemed to be in good spirits, and Alex could detect no animosity. He made a mental note to commend Jones for a difficult job well done.

"Boss, I want you to meet my crew," Jones said as he stepped off the schooner.

"My first Mate, Bruce Yoon." Bruce was medium height, 18 years old, not much muscle, but seemed quite bright.

"My Bos'n Francesca Woodward." Francesca was petite with brightly colored boyish cut hair. She was a cute 17-year-old with a couple of piercings.

"My second mate, Carmina Pebsworth." This 19-year-old stood 165 centimeters, and wore clothing that exposed strategic parts of her anatomy. Her long dirty blond hair parted to expose a nipple peeking through a hole in her top. She grabbed her crotch as she stepped off the schooner.

"And my third mate, Bobby Pfaff. His daddy built this schooner." Bobby was a tattooed, pierced 18-year-old skinhead. He stood 168 centimeters. His eyes appeared dull and lifeless.

"Kids," Jones addressed the motley group, "this is Alex Regent. He's God around here."

Bruce nodded politely, Francesca giggled with her legs spread apart and hands on hips, Carmina grabbed her crotch again, and Bobby gazed vacantly, looking right through Alex.

"Gofer and Jeff, you guys take these kids to the chamber complex and clean them up. Then take them to the galley for some food. I'll meet you there."

Alex grabbed Jones by the elbow and turned him toward the shore. "What do you know about these kids, Jonesy?"

"You're not gonna believe this, Boss, but that little blond chick, Francesca Woodward, is the debutante daughter of the Delaware Woodward banking family." Alex glanced at him. "No shit, Boss. For real." Jones laughed. "She's alright, though. Just a bit weird." Jones laughed again. "The Yoon kid is sharp as a tack. Dad's a millionaire Chinese clothing importer in Los Angeles." Jones lowered his voice a bit. "That Carmina chick—she's sex crazy. Spent most of the trip naked on deck getting a suntan. Got it on right there with Bobby and Francesca—together! Tried to make me, but since I was Skipper, I thought no way!" Jones sighed. "Maybe next time someone else can be skipper." He laughed. "My luck though, they'll all be like Bobby. His elevator don't go all the way to the top. I think he fried his brains on sompin'. Dad's got a shipyard in Connecticut. Builds big ships and little ones—at least that's what I got from the little skinhead. Only time he came alive was when Carmina humped him."

"Thanks, Jonesy. I got it," Alex said. "You did good."

Alex headed for the Main Jarvis Compound atop the berm, where he intended to search out the details of his new charges and find a way to get them home.

CHAPTER ELEVEN

JARVIS ISLAND COMPOUND

Over the next three hours, Alex chased down the families of all four of Jones' "crewmembers."

The Woodwards readily agreed that Francesca's father, Delmer Woodward, would make arrangements to pick her up. Alex suggested that Woodward contact Carleton Montague immediately, since there might still be time to hitch a ride on his *Gulfstream*. Otherwise, Alex told the Woodwards, the *LLI* corporate jet would be leaving Seattle the next morning. *LLI* would transport Mr. Woodward to Jarvis and back with Francesca for $10,000.

The Yoons were equally concerned, but Bruce's father, George Yoon, was on an international flight to Hong Kong. Alex was able to reach him in the First Class section of his Hong Kong Airways flight. He had an unavoidable meeting with the Chinese government trade representative the following morning. In the meantime, he would attempt to make arrangements to get to Howland, but his firm did not have a corporate jet, and he would be unable to get to Honolulu to catch the *LLI* jet during its fueling stop. It wasn't the money; it was the time factor. Alex would not allow Yoon to speak with his son, and insisted that he personally had to pick up Bruce. Yoon agreed to get there as soon as possible, but opined that it might take several days. Alex was left with the impression that George Yoon had more immediate and important things on his mind than his son.

Alex's conversation with Reyes Pebsworth was a failure. At first, Pebsworth called Carmina a "fucking slut," and terminated the Link. Alex initiated an executive override on the second call. Once Pebsworth stopped sputtering and accepted his inability to terminate the Link or continue with what he had been doing before the call, he listened to Alex. The bottom line of their conversation was that Pebsworth would not raise a finger to help his "slut of a daughter." If this is what it took to get her attention, then so-be-it, he told Alex.

About twenty minutes later, Alex received a call from Mindi Pebsworth, Carmina's mother. She was a dazzling forty-five-year-old replica of Carmina. She obviously had taken a few minutes to "freshen up" herself before placing the call. Her simulacrum in Alex's holodisplay oozed charm and motherly concern. She apologized for Reyes' behavior, and pleaded with Alex to let her pick up her daughter. It would be, she explained, her first real opportunity to be with her daughter for an extended period. Perhaps, Mindi pleaded, she could get through to her. Alex found himself unable to resist her pleas and agreed that Mindi would make the pickup. She would meet the *LLI* corporate jet the next morning in Seattle.

Alex's outreach to the Pfaffs was a complete disaster. He never got through to any immediate family member. He did speak with a senior aid who told him that Bobby was disowned and completely on his own. The *Green Avenger* had been purchased at a huge discount by Lars Watson through a dummy charity front. Watson had so arranged it that the Pfaffs thought they were helping Bobby get his feet wet in a major charity. When his father, Johnathan Pfaff, discovered what *Green Avenger* was really being used for, and the apparent role Bobby had played in getting the vessel, he washed his hands of Bobby, and cut him off totally. Apparently, this was the beginning of Bobby's downward spiral into drugs and skinhead rebellion. There was no chance, the aid told Alex, that anybody would come for Bobby. In fact, he said, the Pfaffs would not even be told of Alex's call.

It took several minutes for Alex to think his way through this development. He found it difficult to imagine parents so cold and calculating, so unwilling to yield, even in the face of their son's possible destruction. The more he thought about it, the increasingly angry he became at their callousness. He called Mabel, catching her at her desk in the Smith Tower Penthouse

He briefed her on the developments with the four families. "Mabel, Bruce is a pleasant enough fellow. I don't mind having him around for a few days. Bobby is another matter. With your permission, I would like to dry him out and see if I can reach him. The kid was sufficiently resourceful to pull that stunt with Watson and the *Green Avenger*. If any of that is left, maybe I can reeducate him. You have any problems with that, Mabel?"

Mabel smiled warmly at Alex. "Don't you have enough to do?" she asked.

"Mabel…"

"Okay, Alex, you soft-headed SOB. I'll give you a month. Give me weekly reports. If you make no real progress, I'll send the Federal Marshals for him." She smiled warmly again. "That okay with you?"

Alex just grinned at her. "Any further progress at your end?" he asked.

"Getting close," Mabel answered, "but you have to be patient a bit longer."

JARVIS ISLAND TARMAC

Early the following morning Alex received a SSRS call. "*Floater Two* to Alex. This is Tex. I'm about a half-hour out with two passengers, a Carleton Montague and a Delmer Woodward. I need a nap, but these guys are hot to trot, especially Montague."

"I'll meet you on the tarmac, Tex." Alex ran down his mental checklist. "You hungry, Tex? Want some breakfast?"

"Naw—I jest want some shut-eye."

Alex had already arranged for a private breakfast for five in the executive dining room. Normally, the executive dining room was closed. Alex and the entire management staff on the project ate with the crew whenever they visited outlying areas. It was part of Alex's management style that clearly communicated to each person on the project that every job was vital, and no one was more privileged than anyone else.

Alex wanted the fathers to be with their daughters for a while, supervised by himself, but otherwise free to follow whatever developed. He was looking forward to the meeting—sort of.

A fierce tropical sun beat down from an azure sky. Fluffy white clouds dotted the northern sky, and the western horizon was hidden by a line of squalls. A light tropical breeze blew from the southeast, crossing a swell that pounded the southern beach from the southwest. *Floater Two* rolled to a stop a few meters from where Alex stood with Tiffany and Francesca. Somehow, Tiffany had managed to make herself appear cool and sophisticated, a beautiful, aloof young woman of the world. Francesca was her typical bubbly tom-boy self, cleaned up, but without make-up, happy and excited to see her daddy.

The back door opened, and Delmer Woodward was the first to disembark. Francesca squealed, ran up to him, and threw herself into his arms.

"Daddy! Daddy! I love you, Daddy! I love you, Daddy!" Tears of happiness streamed down her cheeks.

Woodward held her gently to his chest. "Baby, Baby, Daddy's here now. Everything's fine now. Daddy's here."

While this reunion was going on, Carleton Montague stepped down from the cabin. For a moment, he watched the touching reunion, and then he turned to the stunning young woman standing before him.

"Tiffany," he said, almost formally, nodding his head.

"Father." Tiffany appeared equally formal.

Alex watched the exchange, thinking that he might be looking at his first failure with this bunch. He could see the anguish in Montague's eyes, but Tiffany held her emotions in check, looking at her father stoically.

For several seconds they stood thus, about a half-meter apart, looking at each other. Then, Tiffany stepped toward her father and kissed both his cheeks, French style, and stepped back. She started to say something to Montague, and Alex strained to hear her voice carried to him on the breeze. "You came for me, Father—you actually came for me."

Montague's eyes filled with tears. He struggled to retain them and his dignity, but Tiffany was making that difficult as she kissed his forehead. Alex barely heard her quiet "Thank you!"

Then the floodgates burst, and the dignified oil financier broke down as he gripped Tiffany's shoulders. "My little girl, oh my baby little girl, what have they done to you? What have they done to you?" Tears streamed down his face. "You're safe now, Baby, you're safe with me!"

Alex stood back, watching the reunions, and found himself moved to tears as well. *There's a chance,* he thought. *These crusty old bastards really do have a heart.* He turned away to recover his composure.

JARVIS ISLAND COMPOUND

Alex let things settle down for a few more minutes, and then he announced, "It's getting hot out here, folks. Let's go inside to freshen up, for a bite to eat, and some discussion and planning." He gestured to the white pickup behind him. "You two girls sit up front with me so your fathers don't have to fight about who gets to sit next to his daughter," Alex said light-heartedly. Francesca giggled, Tiffany smiled aloofly, and their fathers chuckled. Ten minutes later they were seated at a round nicely set table in the cool, slightly darkened executive dining room.

"Not bad for the boondocks," Montague remarked as they sat down, and he squeezed Tiffany's shoulders.

"This is a significant part of what will become the world's major access to space," Alex said wryly as he joined them, sitting between Francesca and Montague.

They were each handed a formal menu by a uniformed waiter of indeterminate background, but with a hint of South Sea Islander. "Thanks, Jake," Alex said. Jake grinned and winked at the girls. "Jake joined us from New Zealand," Alex said. "He has God knows what in his background, but there is sufficient Maori that he can officially claim Maori citizenship. Nowadays, it's more a matter of pride than anything else, but you have to hear the women in his "native" group sing. It'll make you cry all over again."

At that moment, a melodic wave of bell-like clear female voices singing as one voice swept through the room. The song contained hints of old-time-religion melodies, ancient Hawaiian chants, Tahitian calls, and more, sung in the ancient Maori tongue, unintelligible, and at the same time universally understandable. It was aural magic. The five people at the table were silent as the music swept over them, cleansing their minds and moving their souls.

"Oh my…" Tiffany said quietly. Francesca's eyes got even bigger than they normally were, and a tear dribbled down her cheek. Woodward slipped his arm around his daughter's shoulders and touched a napkin to his eyes. Montague lifted Tiffany's hand to his lips and heaved a deep sigh. Then he smiled at Alex. "You are one hell of a negotiator," he said. "And you don't play fair."

Francesca giggled, Tiffany smiled, and Alex chuckled quietly. The ice was broken.

※

Following a leisurely breakfast filled with small talk about family and friends, talk that Alex listened to rather than participate in, the group retired to the conference room next door. This room was equipped with the same intricate holoprojector that Alex had used to present the project to Lori the day before—although it seemed like a month ago.

Alex walked them through the concept, starting with a garden hose squirting water that can elevate itself without external support. He showed them how they had converted Howland Island from a desolate tropical rock to a thriving hub of eco-friendly commerce. He walked them through the process of constructing Slingshot, showed them how the Eastern and Western Complexes generated power, how they enhanced the local environment, and how the entire massive project was the most ecologically benign large engineering project in the history of humankind. He showed them how the rail would be elevated, and how the entire launch loop would operate. He inserted simulations of future capsules climbing the elevator to *Earhart Skyport*, being launched, and ultimately being flung on a trajectory to Mars by a spidery tether at the L-4 LaGrange point.

As he talked, Alex emphasized again and again how this was the very first major project that could be supported by nearly every point of view. Then he detailed the damage caused by Lars Watson and his *Green Force* crew, including Tiffany and Francesca. Tiffany was appalled and began sobbing softly onto her father's shoulder. Francesca's large eyes filled with sober tears. She lost her bubbly demeanor, and mouthed softly to Alex, "I'm so sorry—so sorry."

Montague and Woodward looked at each other, seeming to communicate wordlessly. They nodded at each other, and Woodward indicated for Montague to take the lead.

Montague folded his hands on the table, cleared his throat, and said to Alex, "How much did these shenanigans cost you, Mr. Regent?" His voice contained a full measure of emotion.

"Call me Alex, please." Alex smiled at each man. "It's difficult to measure, but my accountants tell me it cost at least a million, maybe twice that." Other than a lifted eyebrow from Montague, there was no reaction from the men, although the girls' faces dropped in shame.

"How can we help?" Woodward asked.

"I don't know where to begin," Alex said. "We need a couple more floatplanes with pilots, a couple high-speed skimmers. We're short on service personnel—Jake and his counterpart on Baker could use twice as many people. We need multi-tasking engineers who can work underwater as well as on the surface. We're lean and mean out here—perhaps a bit too much so." Alex spread his hands.

The two men looked at each other again. "Okay—you got the planes and skimmers, and drivers for both," Montague said. "Speaking for myself," he continued, "I had no idea what you were doing out here. This is incredible, unbelievable."

Woodward interrupted, "I concur. I—we—want to be a part of this." Montague nodded.

"I know you already spoke with Mabel," Alex said to Montague.

"I spoke with her, as well," said Woodward.

"We'll handle the Mabel end," Montague said. "We'll get the equipment to you as soon as possible." He squeezed Tiffany, who had lost some of her aloofness.

Woodward slid his chair back, stood up, and pulled Francesca to her feet. "Come here, Baby," he said as he took his daughter into his arms. "Welcome home."

"You gents ready to meet Mr. Watson?" Alex asked. Both girls looked startled. "Don't worry," Alex told them. "Jake," who had just appeared from the galley, "will take you for an interesting tour of Jarvis while your dads and I take care of this business." The girls nodded and left with Jake. Francesca's bubbly demeanor had returned, and she skipped out, chattering as they left.

"Let's go meet Lars," Alex said as the three men departed the dining room. Alex asked them to walk with him so he could brief them. As the men walked the ten-minute distance to the chamber complex where Watson was stowed, he told them what he knew about Watson.

"He's like one of those fanatical religious leaders," Alex told them. "Apparently, none of his followers are very science savvy. He feeds them a line of bull in flowery scientific-sounding terms, and they swallow it hook, line, and sinker." Alex grimaced. "He gets them to do his dirty work, although when the action is critical, he's there with them, directing them, and often doing the dangerous stuff himself." Alex kicked a coral pebble off the road. "Funny thing is, he actually seems to care for them. Trouble is," Alex sighed, "the girls seem to reciprocate. I suspect he gets it on with every female in his immediate group." The two men stopped and turned to Alex, who was walking between them. "I'm sorry," Alex shook his head, "but you're about to meet the scoundrel who seduced both your daughters." Alex faced them both. "I wish it were different, but," he shrugged his shoulders with hands held out, "that's apparently how it is."

Both fathers reacted with obvious anger, but Montague's was palpable. His oilman pedigree was just under the surface. He was larger and tougher than Woodward, who looked like a typical banker. "Goddammit!" Montague pounded a fist into his palm. "Goddammit it to hell!"

Woodward said nothing, but his face took on a fiercely determined look.

Shortly thereafter, the three men entered the chamber complex, and Alex led them down a hall to a locked door. "Are we ready?" Alex said.

"Unlock the door," Montague growled.

"Please," Woodward added.

Alex unlocked and opened the door. Watson was sitting in the straight-backed chair, hands folded on the table in front of him.

"On your goddamn feet," Montague roared, pushing into the room.

"Easy," Alex cautioned, placing his hand on Montague's shoulder.

Watson pushed himself back, knocking his chair over backward. Montague took a deep breath. "What did you do to my little girl?" he said through clenched teeth.

"Lars Watson," Alex said, "meet Tiffany's father."

"Oh shit," Watson mumbled, backing away.

"And meet Francesca's father too." Alex indicated Woodward. Woodward glowered at Watson, who shrank further back.

"I'll leave you gentlemen to get acquainted," Alex said with a grin as he left the room.

A half-hour later, Alex returned and opened the door. Watson was sitting at the table glumly silent. His left eye was black, his right cheekbone was badly bruised, and his bottom lip was split open. Montague was nursing the knuckles of his right hand.

"Remind me not to piss off Carleton here," Woodward said softly to Alex. "That guy packs quite a punch."

"Are we old friends now?" Alex asked cheerfully.

"Fuck you!" Watson mumbled through bruised lips.

"Can we talk for a minute?" Montague asked Alex, putting his hand on Woodward's shoulder.

"Sure. Let's step outside."

Alex locked the door, and they went into the chamber room down the hall.

"Let me—us—take that bastard off your hands," Montague said. Woodward nodded.

Alex raised an eyebrow.

"No, we won't kill him, but he'll wish we had," Montague said.

"No problem," Alex said. *That's one more problem eliminated*, he thought as they walked back to the main compound.

※

Two days later, under a high stratospheric haze punctuated with wispy clouds carried to the west by upper-level winds, Sanchez announced that he would be landing with Mindi Pebsworth in tow. "Look out for this one," he said. "She's got claws."

Mrs. Pebsworth turned out to be every bit as charming and beautiful as she had appeared on his Link. She was warm, slightly sexy, and infinitely grateful for being allowed to get her daughter. "Please call me Mindi," she insisted.

Carmina, on the other hand, actually spat at her mother when they met. Alex handed Mindi a handkerchief to wipe her face while Carmina favored him with a middle finger. He was very impressed by how Mindi handled it.

"Please, Carmina Dear, let's give each other a chance." She reached out and took Carmina's hand. To Alex's surprise, Carmina didn't pull her hand away, and she didn't grab her crotch either, he thought.

"Walk with me," Mindi said to her reluctant daughter. "I want to show you something."

As the late morning tropical sun beat down on them, they strolled toward the main Jarvis Compound. Wispy clouds seemed to fade in and out in the blue overhead dome as they whipped westward, but they offered no relief from the ever-present sun. Alex followed at a respectful distance, but still sufficiently close to hear.

Mindi reached into her shoulder bag and extracted a small holoprojector. She made a couple of adjustments to the small device, and a shimmering appeared in the air about a foot in front of mother and daughter. The shimmering coalesced into a still holo of a very pretty girl who could have been Carmina, except Mindi said it was a holo of herself when she was Carmina's age. Despite the bright sun, the holo was sharp and clear. The girl was dressed similarly to Carmina, the same outrageous flaunting of convention, the same raunchy attitude toward sex—for Alex, it was really difficult to tell them apart.

"We're not very different, you and I," Mindi said to her daughter. "You are so much like me at your age that it scares me—because I know what I did and the chances I took." As she talked, the holos changed from one to another, presenting a mélange of different views of a very pretty, but very mixed-up girl. "If your father had not rescued me, I don't know what would have happened. For certain," she reached out an arm and squeezed her daughter, but then dropped her arm again, "you would not have happened. I probably would have ended up in some third-world brothel somewhere, and finally probably on the street." Mindi stopped walking and placed herself in front of her daughter. "Carmina," she said gently, "but for your father…" Her voice trailed off.

"Let's go inside, out of the heat," Alex suggested.

They both agreed, and Alex led them to the executive dining room. Jake brought out three cool, tropical fruit juice refreshments.

"I'm here," Mindi said with a quaver in her voice, "to make things up to you." She turned to Alex. "Thank you, Mr. Regent, for making this moment possible."

Alex simply smiled at them.

"Please come back with me," Mindi said to Carmina. "I'm not asking for any promises, except that you'll try. For my part," she turned and smiled at Alex, "with Mr. Regent as my witness, I promise to be the mother I should have been all along." There was a catch in her voice, but she retained control. "You'll be free to come and go as you wish." She reached out for Carmina's hand, but Carmina pulled her hand back and glared at Alex. "Just give both of us this chance." Her voice cracked, her eyes brimmed, and a lone tear rolled down her exquisite cheek. "Please…"

Alex saw Carmina's determination waiver for a moment, and then he saw a sneer rise to her lips. "Why should I, Mindi. You did just fine without your folks…"

Mindi interrupted to say, "That's not so, Dear. Your daddy pulled me away from the brink. I'm only here because of him…"

"Lars…" Carmina started to say.

"Honey, Lars Watson can't hold a candle to your father. He'll use you and throw you out. Trust me, I've been there." Mindi paused and then carried on with her story. "I was the lover of…" She then related a shocking tale of how she had joined an environmental activist group that was a predecessor to *Green Force*. She thought the leader was a prophet of the new age, and she gave herself to him totally. Even when she discovered that he was taking sexual advantage of every other girl in the group, she still held on to him, still believed in him. When he asked her to have sex with specific individuals outside the group, she willingly complied, because it was for the cause, for a better, greener Earth. One day, she found herself in a hotel room with an older man who was somehow different than all the others. He seemed surprised that she was there, stating that he had expected to meet with the leader to discuss a large donation to the movement. Mindi explained to him that she was the leader's incentive for the largest possible donation. At that, the man became angry. He began to ask Mindi about her own background, asked about herself, about her dreams and goals. An hour went by with nothing but conversation. When Mindi finally approached him with bared breasts and lips waiting to be kissed, he gently wrapped his jacket around her shoulders and whispered into her ear that if she would trust him, he would forever remove her from this den of iniquity. She nodded, Mindi said, and a month later, they were married.

"Your father was my savior as sure as if he had climbed down off a cross to rescue me." She paused. "Now, do you understand?" Carmina just stared at her.

"Lars has it right, Mister!" she sneered at Alex, ignoring her mother.

The following morning, accompanied by a relentless tropical sun, Sanchez flew them back to Howland. It wasn't all peaches and cream by any means. But Alex considered it a definite start. He was willing to put his money on Mindi, he related to Margo by Link after they left.

JARVIS ISLAND—SOUTHERN BEACH

The next day Alex and Bruce were walking together along the southern beach. Off to their right white-crested waves crashed onto the glistening white-coral sand beach. Occasional rusting iron braces, still remaining from the *Amaranth* wreck in 1913, protruded through the pristine sand, oddly out of place on this tropical beach. From time to time, the surf uncovered a rounded lump of coal, mute evidence of *Amaranth's* cargo. Bruce had requested the meeting, and since Alex needed to speak with Bruce anyway, it was a fortuitous stroll.

"You talk with my father, Mr. Regent?" Bruce spoke with a polite tone.

"Yes, Bruce, I did." Alex told him about his finally reaching George Yoon on his international flight, and about Yoon's complicated business affairs in Hong Kong.

"That's my dad." Bruce chuckled shyly. "He's a good man, Mr. Regent, and he cares a great deal about me. I'm afraid I've been a big disappointment to him. But he'll be here as soon as he can. You'll see."

"That's kinda why I wanted to speak with you, Bruce." Alex sat down on the raised cut jutting above the rough coral sand, and patted the ground beside him. Bruce joined him.

"Your father will be here in a couple of days. I spoke with him this morning." Alex leaned back with his hands holding his right knee. "He cannot understand what you are doing." Alex smiled. "To say he's bewildered would be a real understatement." Alex waited for a full thirty seconds. "So tell me, Bruce, what's really going on?"

"That's why I asked to meet with you, Sir." Bruce hesitated, his shyness nearly overcoming him. "You see, Mr. Regent, I'm not exactly what you think I am."

"Okay, I'll bite. What are you talking about." Alex was genuinely curious.

"I'm here undercover."

"Say what?"

"I'm here undercover. I am conducting an undergraduate journalism project. I joined *Green Force* to conduct some in-depth investigative journalism research." He grinned shyly at Alex. "I give you my word, Mr. Regent. I did nothing to sabotage Slingshot." He got up and walked a few meters toward the water, and stooped to pick up a lump of coal. He turned back, tossing the lump from hand to hand. "I could hardly stop the others without revealing my subterfuge, but I promise you, Sir, I did not help them." Bruce sat down again next to Alex and continued conspiratorially, "You should read my notes. You would be astonished at what these idiots are doing."

Alex had a difficult time restraining his astonishment. He had recognized that Bruce was somehow different than the others, but he had completely missed this twist. He made up his mind almost instantly.

"Bruce," Alex said, "how would you like a job?"

Bruce's reaction amused Alex. He stood, opening and closing his mouth, but otherwise speechless. Finally, he asked, "But what about the rest of my schooling?"

Alex stood up and approached Bruce. "I need someone to keep tabs on the media, to compile routine reports to the media, and to liaison with the media. I know you are just getting your feet wet as a journalist, but this is an opportunity that rarely comes along in a journalist's career, and never for one so young as you. My corporate staff will arrange with UCLA for you to attend your remaining classes holographically. We'll pick up the tab for your remaining school—including graduate school." Alex reached out and gripped Bruce's shoulder. "All you have to do," he added, "is confirm my judgment. Think you can handle that?" Alex's eyes twinkled.

Bruce was speechless, but nodded his head, and nodded his head, and nodded his head, as they walked back to the Jarvis Compound. As they entered the building some fifteen minutes later, Bruce finally found the words.

"I accept," he said.

CHAPTER TWELVE

JARVIS ISLAND COMPOUND

George Yoon arrived at Jarvis in *Floater Two* after a bumpy ride. Sanchez told Yoon when he picked him up at Howland that weather was closing in, and that nobody would blame him if he waited another day at Howland.

"I'm always putting Bruce second place to business," Yoon told Sanchez, "but not this time." He swept his eyes across the angry sky. "How safe is it?"

"It's not as bad as it looks," Sanchez said. "If we leave right away, we'll avoid the worst of it." He reached for Yoon's case. "With a bit of luck, we'll be there in about three hours."

Sanchez and his passenger dodged thunderheads and negotiated shears for a thousand nautical miles. They landed on Jarvis in the late afternoon in a driving rain, with the wind picking up even as they landed. Yoon was determined to see Bruce as soon as possible. Sanchez escorted him by pickup to the Jarvis Compound as the stormy sky darkened into early evening. Off to the northeast, lightning flashes illuminated the low overcast as the water roiled against the northern shore.

Alex and Bruce met Yoon in the Executive Dining Room. Father and son did not hug, but they gripped each other's forearms warmly. They sat at the round table, and Jake served a light meal, during which Alex described Slingshot to Yoon.

Over coffee and mangoes in cream sauce, Alex suggested to Bruce that he come clean with his father, and tell him about his undercover role with *Green Force* for the past few months.

Haltingly, Bruce began his tale, finally finishing with, "So you see, Father, your faith in me was justified after all."

"Mr. Regent," Yoon said, "I cannot thank you enough for bringing us together. And Bruce," he said, turning toward his son, "we have some catching up to do."

"Father, there's more." Then Bruce told his father about the job offer from Alex, and what it would mean to his future as a journalist.

Yoon's reaction was surprise and genuine pride. "You'll keep in touch, won't you?" he said to Bruce.

Bruce's answer was drowned out by crashing thunder as the storm passed over the island. Alex shook Yoon's hand, and Jake showed father and son to their quarters for the night.

※

Sanchez had planned to continue on to the Eastern Complex to get Lori and Dex, but he let discretion overtake valor, and decided to wait out the storm. He joined the rest of the crew in the cafeteria. As he entered, he looked around the filled tables for Jeff, Gofer, and Jonesy. He spotted them in a corner, along with Carey and Bobby from *Green Force*.

"Hey, guys," he said as he sauntered to their table. "Got room for a damn—I mean damp—aviator?" Chairs moved, another appeared, and Sanchez took a deep drag on the beer that appeared before him. "I needed that," he said with a sigh and a smile. "So, what's happening?" He had genuinely been too busy to keep track of the comings and goings of the *Green Force* gang, except for his part in ferrying parents and charges to and from Jarvis.

"Tex, meet the newest member of the team," Jones said, patting Carey's back. "The boss hired him as a nerdly." Carey blushed, and everybody laughed.

"No shit, Jonesy," Sanchez said, reaching out to shake Carey's hand. "How d'ju pull that off, Kid? I thought for sure you'd be somebody's Bitch in the pen."

"Don't know for sure," Carey said, his voice breaking into a squeak. "One moment, I'm for sure on my way to the pen, and the next, the Boss is saying he admired my hacking skills, and wants to know if I want a job. I mean, like, once he explained to me how you guys are the greenest ever, and Slingshot here, is the greenest ever, and how maybe Lars is missing the point…" Carey looked around the group, and then straight at Sanchez with a shy smile. "Well, you know," he squeaked, "I thought maybe I could, like, lend a hand."

"Hey, Bobby, what's up?" Sanchez said with a grin to the little skinhead quietly staring at the tabletop in front of him.

"Leave him alone, Tex," Jones said, in his heavy Philly accent. "Bobby here's working through withdrawal. Alex put him on cold turkey. He's kinda hurtin'." Jones turned to Bobby and continued, "Bobby, we'll get you cleaned up, Kid. Just hang in there for a bit."

If Bobby responded at all, it was a grunt. Sanchez raised his bottle in a quiet toast to Bobby's travail.

EASTERN COMPLEX

Early the next morning, before the sun had a chance to heat the tarmac to frying-pan hot, Sanchez popped an *Alim*, to eliminate any vestiges of his beer-drinking relaxation of the previous night, and revved up *Floater Two*.

"Eastern Complex, it's Tex. Be there in about an hour."

"It's Stu, Tex. Roger that. We're ready for you. The press is chomping at the bit."

Sanchez quickly gained altitude as he lifted into the morning sun. He could just make out a few bright stars above the western horizon that had not yet been washed out by the ball of tropical fire ahead of him. The only reminder of last night's storm was the unsettled sea surface below him. His sonic projectors had driven the birds away from his immediate vicinity, but as he lifted his gaze from the dimming stars behind him, he watched thousands of birds on stretched wing riding the turbulence he left in his wake. It never ceased to amaze him how these magnificent birds seemed to disappear during a storm, and then reappear in even greater numbers in what seemed like a celebration to the return of normalcy.

When Sanchez had first arrived at the Eastern Complex, and had watched the massive construction on Jarvis, he wondered what would happen to the several million birds that made the island their home. As it turned out, he needn't have worried since birds and humans formed a genuine symbiosis. This was nowhere more true than at the Eastern Complex, where the OTEC well significantly increased the available food, resulting in an even larger bird population than before the project started.

Jarvis quickly disappeared behind him as Sanchez pushed his twin-engine floater to its limits. These marvelous little twin-turboprops really made the twenty-five-hundred-klick stretch of Slingshot seem like a community. All the more so, of course, considering that the nearest real civilization lay twenty-five hundred klicks to the north—Honolulu.

Sanchez arrived over the Eastern Complex not long after the sun had fully formed to the east. His arrival signaled the rise of thousands of birds that had made a semi-permanent home around the complex. He triggered his sonic projectors to prevent sucking a bird into his intakes and dropped down to the mirror-like surface shielded by the OTEC well. Stewart, Lori, and Dex met him as he glided to the dock.

"You guys are gonna let a man get a bite to eat before we leave, right?" Sanchez shook the men's hands and hugged Lori. "Load yer gear any way you want, and meet me in the galley." And turning to Stewart, "Your guys'll fuel it, right?" He left the three to load their equipment and fuel the plane while he dropped through the pressure hatch and found his way to the cafeteria.

Sanchez grabbed a mug of coffee, ordered a plate of *huevos rancheros*, and in short order, was relaxing over a home-cooked meal that was as good as anything he could get back home in Texas. "Don't forget the Tabasco," he had told the duty mess-cook with a wide grin.

A few minutes later, Stewart, Lori, and Dex joined Sanchez. "We've already eaten," Stewart said, "but we'll have coffee."

Dex took the opportunity to record the scene for editorial insertion at an opportune point in the eventual documentary.

Sanchez brought Lori up to date on the happenings with the *Green Force* people. "So Tiffany, Francesca, and Carmina are gone," he said. "Too bad you couldn't have met Francesca," he told her. "She was a little doll."

Sanchez finished his breakfast and stood up. "Whenever you're ready," he said. "I'm gonna get rid of some of this coffee, and then I'll meet you at the floater."

"I'd better," Lori turned to Stewart and giggled, "what did you call it? Pump my bilges?" He nodded with a chuckle. "There's no lady's room on that small plane," she said as she left the room.

JARVIS ISLAND

Island for the benefit of his excited passengers. The sea was running from west of southwest, long rollers about fifty meters apart. From 3,000 meters Lori and Dex could clearly see how the waves refracted around the island, so that the western, northern, and southern beaches were awash with white foaming breakers spending their energy against the coral sand. Directly to the west, the artificial harbor formed by the floating breakwaters presented a nearly calm surface despite the high-energy rollers crashing against the floating structures. Glistening brightly against the gray-green background of the prevailing vegetation, the Jarvis Compound was centered on the northern berm, the dive facility on the southwestern berm, and the socket still under construction at the island center; all stood out dressed in blinding coral-white. Birds were everywhere—sooty terns, masked boobies, lesser frigatebirds. They filled the sky like a swarm of overgrown mosquitoes, except they glided gracefully on outstretched wing, picking up every thermal, every eddy. Occasionally one would drop to the sea surface to collect a prize, and the terns and boobies kept an eye out for the ever-present frigatebird attempting to steal whatever they had just captured.

Dropping lower while he circled, Sanchez finally flared over the tarmac and rolled to a stop near the Jarvis Compound. Alex stood alone on the tarmac, a white pickup behind him. A couple of his men waited with the truck. Lori was the first out of the floater.

She ran to Alex, threw her arms around his neck, and kissed him. She pressed her body against him in a very non-sisterly way, and Alex made an executive decision to give a little bit back, and then to disengage himself.

"Hi, Lori," he said, and then turned to shake Dex's hand. "That," he said, "will be erased. Do I have your word?"

"No problem, Alex. Since I didn't record it, no problem."

The Jarvis crew loaded the equipment and themselves into the pickup bed. Alex drove the short distance to the Jarvis Compound, where everyone unloaded, and Jake showed the visitors to their rooms.

An hour or so later, Alex assembled the group in the executive dining room for lunch and a briefing about the *Green Avenger* and her motley crew.

"There's nothing going on at Jarvis that justifies a live holocast," Alex told them, "but I will be happy to give you access to the *Green Avenger* crewmembers who still are with us, and to the rest of my people here so you can put together a good show about the sabotage, and what role Jarvis Island plays in this project."

With that, Alex launched into a description of their encounter with *Green Avenger* along the submerged double rail. He described how they handled the situation, leaving out the specific technique they used to bring the divers to the surface. He described Watson and each of his team members and related how he had contacted their families. Then he told them how Montague, Woodward, and Mindi had flown in to retrieve their children, and related with a chuckle how the two men took Watson back with them. Then Alex introduced George Yoon and Bruce. He let Bruce tell his story in his own words, and made sure that Lori completely understood that Bruce was his new liaison with the media in general, and with them in particular.

From Alex's perspective, this allowed him to get Bruce into the action immediately, and it put a space between himself and Lori.

They finished their lunch, and Lori and Dex spent another hour interviewing Jones, Carey, Bruce, and his father. In the meantime, Alex cornered a reluctant Bobby and set out for a walk around the island.

JARVIS ISLAND—SOUTHERN BEACH

The sun had passed its zenith in the azure dome, but still retained most of its tropical impact as Alex and Bobby worked their way around the southwest corner of Jarvis and commenced walking on the glistening white-coral sand beach. The tide was out, and most of the reef was exposed, extending about a hundred meters from the sand. The rollers left over from the storm broke against the outer edge, dumping their energy and excess water onto the reef, and flowing toward the sand, gurgling and bubbling over the living coral, and finally disappearing before actually reaching the sand. Every once in a while, an especially large roller surged sufficiently far to dampen one or two meters of sand.

"Did you notice," Alex said, "that every seventh to tenth roller reaches the sand?" Alex could tell that Bobby had started counting the incoming rollers. He intended to keep matters light, but he was determined to reach the boy—if reaching still was possible.

The little skinhead was dressed in ragged shorts of indeterminate color and an open faded button-up shirt without sleeves. Laceless sneakers protected his feet from the burning sand. His unprotected shaved pate glistened with perspiration. Alex reached into his pocket and handed Bobby a kerchief. "Let's four-corner knot this to protect your scalp," Alex told him. "You've got enough problems without adding heat stroke."

Bobby actually smiled and thanked Alex. Between them, they managed to effect a reasonably efficient scalp cover, and then they continued their walk.

Bobby's gaze shifted from the incoming rollers to the iron ties protruding through the sand and the occasional rounded clump of anthracite sitting atop the sand, contrasting sharply with the glistening white.

"Know what those are, Bobby?" Alex asked casually.

"Shipwreck of some kind, I guess," Bobby answered, almost as if the words escaped his lips reluctantly.

"I guess you'd know with your background," Alex said. "Let me tell you about it."

As they strolled along, stopping occasionally to examine a piece of old wreckage, Alex told Bobby the story of the wreck of the *Amaranth*.

"It was August 30, 1913, just past eight in the evening," Alex said. "The barquentine *Amaranth* was on her way to San Francisco from Newcastle, New South Wales, with a cargo of coal." Bobby looked up at Alex with a question on his face. "Australia, Kid. That's about a hundred fifty klicks up the coast from Sydney." Alex grinned at him and received a tentative smile in return. "Anyway, Ship's Master C. W. Neilson was running the *Amaranth* under full sail on a northeast track. The weather was slightly foggy, but his track lay well to the west of Jarvis Island, and the Master was unworried, since the track was virtually open all the way to San Francisco.

"Suddenly, without warning, the *Amaranth* struck that reef over there," Alex pointed to a spot just to the west where the reef jutted out another hundred and fifty meters or so, "and heeled over. According to the Master's log, the sea was nearly calm, and there was no surf at all. Even after they struck, they could see nothing of reef or land. The crew and passengers took to the boats, and stood by the heeled-over wreck until morning. The next morning they headed around the western side of the low-lying island and landed near where the wharf is inside our artificial harbor. They found the remains of old buildings and identified a few graves," Alex turned and pointed back the way they had come, "there over the berm. We'll look at them later. Unfortunately, there was neither sign of life nor useful vegetation—thousands upon thousands of birds, but that was it."

As Alex said this, as if on cue, birds rose from all over the island, filling the sky with their skreeghing, swooping, and soaring with a cacophony of visual and aural stimulation. Then they watched the birds part as a floater came in for a landing. Alex didn't recognize the tail markings. "Looks like we got a new floater," Alex told Bobby. "I think it's from Tiffany's and Francesca's fathers." He winked at Bobby. "They're on our side now, you know." They turned and continued their walk as Alex picked up his tale.

"Anyway, as the party landed, they saw a big sign that read: *This island is leased by His Britannic Majesty King George to the Pacific Phosphate Company of London and Melbourne. All trespassers will be prosecuted under English rules.*"

"That's absolutely crazy," Bobby said, with obvious interest, much to Alex's delight.

"It didn't take long," Alex continued, "for *Amaranth* to break up. It was continuously pounded by waves, just like the ones you see out there right now. The calm on the night they hit the reef was a definite exception. The poop deck remained exposed through the next day, and the crew could retrieve water and provisions. The Captain determined that they had wrecked on Jarvis Island, which was improperly plotted on his chart. Since rescue from Jarvis was virtually impossible, they decided to sail their two lifeboats south to Samoa, some nineteen hundred klicks away.

Image 8—Wreck of the barquentine Amaranth hard aground off the southern coast of Jarvis Island

"The Captain's boat included Mrs. Neilsen and their 18-month-old boy, and six other persons. The second boat was commanded by First Officer A. M. Johnson, and held six people. Two days after the wreck, they set out on the morning of September 1. They rigged masts from oars and set sail for Samoa. Two days out, they parted company to increase their odds of survival. On September 11, the Captain's boat reached Pukapuka, at the northern end of the Cook Islands, sixteen hundred klicks from Jarvis, where the Polynesian islanders treated them kindly. As ships did not normally stop there, they sailed on another four days to Pago Pago, five-hundred-fifty klicks further south, where they were put up in natives' barracks. Fortunately, the US Gunboat *Princeton* was stationed there, and they received some clean clothing from the sailors. *Princeton* commenced a search for the other lifeboat and found it on September 24 at Apia, some two hundred klicks to the northwest on the island of Upolo. The crewmembers had run out of water but were fine otherwise. Several days later, in Pago Pago everyone boarded the *SS Ventura*, bound for San Francisco via Honolulu.

"A crewmember named Vining told reporters in Honolulu that some of the crew had suffered from mild scurvy. He reported that, despite long dreary days and nights without seeing a sail, and with occasional buffeting by winds and rough seas, Mrs. Neilsen and her child had remained cheerful. 'The baby had been a wonderful kid,' he said, 'ready to laugh and crow at all times. He had kept all of us in good humor.'

"And that," Alex said, "is the story of the wreck of the *Amaranth*."

They walked on in silence for a while, and then Bobby said, "Mr. Regent, what's gonna happen to me?"

"That's up to you, Bobby." As they walked on, Alex continued, "How are you feeling now? You know…the drugs and all?"

"I could use a toke…"

"Out of the question, Bobby…"

"I know that, Mr. Regent. Seriously, what are you gonna do with me?" Bobby stopped walking and actually looked Alex in the eyes.

"Are you ready to give up that *Green Force* crap?" Alex asked.

"I dunno…it seems so important."

As they continued walking, Alex patiently stepped Bobby through the same arguments he had used earlier on Carey and the girls. "You've been had, Bobby. It's as simple as that," Alex said. "You're a naïve kid, and Lars took advantage of you." Alex put an arm around the boy's shoulder. "You've been had, you lost your family, and now you face God knows how much prison time." They walked in silence for a while, rounding the southeastern point, where the reef was still uncovered, but water was beginning to boil up through the reef with the incoming tide. Overhead, afternoon clouds partially filled the northwestern sky.

"I got no place to go, Mr. Regent…" Bobby's voice trailed off. "No place…"

"What are your skills, Bobby?"

"Nothin' really, Sir. I, like, dropped out of Yale and ran off with Lars." He paused. "I guess I, like, know how to sail and run a boat, you know, things like that. I guess I'm, like, pretty good at that." Bobby was hesitant, feeling his way through uncharted waters.

They walked on in silence around the tiny island, a man at the very top of his profession and a boy who had totally lost his way. As they walked, stopping from time to time to examine a rock, a shell, or a piece of flotsam and jetsam, a bond began to grow between them, tenuous to be sure, but there nonetheless.

They returned to the Jarvis Compound an hour or so before dinner. "Get cleaned up and join me and the others," Alex said. "I have to check on the new floater."

CHAPTER THIRTEEN

SEATTLE—SMITH TOWER

In 1914, in what then was downtown Seattle, firearm and typewriter magnate L.C. Smith built a tower overlooking Pioneer Square. It was accurately touted as the tallest building in the world outside of Manhattan. In 1976 it received its first exterior cleaning, and in 1999 its first updating to then-modern standards. More recently, it was completely renovated, with the top four floors being occupied by Launch Loop International LLC. Floor 35 was originally the famous Chinese Room, used for special occasions, including the marriage reception of Smith's granddaughter. Now it housed the main offices of LLI. The top three floors had been converted into a penthouse condo in 1999. Now they housed the private offices of the *LLI* Board Chairman, Mabel Fitzwinters.

Mabel was cast in the mold of a previous female titan of the scientific and political worlds of the twentieth century, Dixie Lee Ray. Dr. Ray was a professor of Marine Biology at the University of Washington, Director of the Pacific Science Center, the first woman Director of the Atomic Energy Commission, and the first woman governor of Washington State. Like Dr. Ray, Mabel was a large woman who learned early in life that she would not go very far on her looks. Fortunately, Mabel was blessed with a powerful intellect. Early on, she discovered two books written by Dr. Ray, *Environmental Overkill* and *Trashing the Planet*. These fascinating volumes became the guiding force in her life. She developed her critical thinking while still quite young, quickly learning to distinguish between real science and junk science that masqueraded as real science in many modern centers of higher learning. Like Dr. Ray, Mabel received her PhD at Stanford University, but she chose Applied Science and Technology, and she followed her PhD with an MBA from the University of Chicago—Department of Economics.

Mabel developed an early interest in space flight, and her PhD dissertation examined the effectiveness of various methods for lifting large amounts of material off-planet. Her initial focus was on hybrid jet and rocket-propelled heavy-lift vehicles. She cut her teeth with a farsighted startup in southern California's Antelope Valley that designed and built innovative launch vehicles and specialized rocket engines. Then she discovered a relatively small group out of Oregon that was working to develop the launch loop concept originally proposed by Keith Lofstrom back in the 1980s. Using her considerable political clout and charm, Mabel was able to talk an aging Pacific Northwest computer software giant into looking more closely at the launch loop concept. She initially got his attention by pointing out that several of his Silicon Valley competitors had thrown in with, and provided considerable funding to, the rocket boys in Antelope Valley.

Although NASA had played around with the concept of a space elevator, it really never had put any value in the launch loop concept itself. Then, with the demise of NASA's manned space program following the disastrous Obama years, space enthusiasts turned their attention to the private sector. Virtually overnight, privately funded satellites found their way into low Earth orbit atop commercial vehicles, followed by many government payloads as well. Then Mabel's old firm put a package into high orbit, and within a few years, the rocket firms were pulling the kind of profits that belonged to the Silicon Valley and Redmond firms of decades earlier. But it was expensive, and the so-called rocket conglomerate established Galaxy Ventures, headed by Sir Justin Crandon of Great Britain, an international cartel with a virtual stranglehold on space commerce. The time was ripe for the launch loop. Under Mabel's keen guidance, *LLI* pulled together the best scientific and engineering minds available. In three short years, she had proved the concept with a working launch loop the size of the new Mariner's stadium in Seattle. *LLI* purchased the old Boeing facility north of Seattle. Mabel hired 3,000 engineers, skilled machinists, and other experts of all kinds. Within three months, production lines were up and running, producing tens of thousands of iron pipe sections, millions of magnets, kilometers of sheath, and all the other myriad elements that went into building a 5,000-kilometer rail that circulated around a 2,500-kilometer-long structure.

Mabel had put together her onsite team consisting of Alex, Margo, and Klaus, and the four of them attacked the halls of Congress with vigor under the guidance of Rex Johnson. Their triumph culminated with the opening of Howland, Baker, and Jarvis Islands to economic exploitation. Within six months, under Alex's able leadership, the three islands were transformed. Howland became the *Amelia Mary Earhart International Airport* with the in situ headquarters for the launch loop, now officially called Slingshot. Within a year, Margo had the in-place ribbon construction underway, Klaus had the OTEC generator construction underway, and the Eastern and Western Complex infrastructures were nearly done. At any given time, Alex had about a thousand workers employed around Slingshot, far less than the ten thousand that now made up the remaining workforce in the Seattle area and several other locations worldwide. Although his operation was spread across twenty-five hundred kilometers of open ocean, it took surprisingly few people to keep the project moving forward as fast as such a thing could be accomplished. Alex spent a lot of time in the air, a lot of time in holoconference, more face-to-face than he liked given his hectic schedule, and what time remained in his offices overlooking Howland's southern reef.

Mabel kept watch over it all from her office high atop Seattle's Smith Tower, where she sat, reviewing urgent communications from across her empire, shifting from Link to Link, the holographic images shimmering above her desk hardly having time to coalesce before being replaced by the next image. While she absorbed the latest information about the pending machinists' strike in Seattle, her Link flashed a distinctive red, and a chime sounded that signaled an incoming encrypted emergency communication. A simulacrum of Carleton Montague appeared in the middle of her holodisplay, blanking the labor report she had been studying.

"Carleton," Mabel said without preamble, nodding at Montague's image. "Why this urgent intrusion?"

"Mabel..." Montague had quickly learned that Mabel was all business, rarely tolerating small talk. "Watson is free," Montague gave her a wry grin. "His legal people obtained a court order forcing us to turn him over to the local authorities, and then immediately filed a *Writ of Habeas Corpus*, and had him on the street an hour later. My guys followed him out of the courthouse, but he had already been whisked off in a chopper that was standing by in the street, if you can imagine something like that. It took us only ten minutes to get our own chopper airborne, but they had already dropped him off somewhere in Jersey. Bottom line, Mabel, we have no idea where he is."

"Thanks for making my day, Carleton."

"There's more, Mabel."

"Check your building roster. You'll find a fourteenth-floor tenant called *Environment, Inc.*" Montague smiled ruefully at Mabel. "That's a dba of *Green Force*." Montague took a deep breath. "They're right under your nose, Mabel. No telling what they are doing, but I would check every line, every circuit, every Link going into and out of your offices. These guys have some big-time backing, and it has nothing to do with the environment."

"You got something I can use, or just idle speculation?"

"I'm working on it, Mabel. Think international—maybe Asian. I'm suggesting that you put your best people on it." Montague disconnected the emergency Link.

Mabel sat quietly, pondering the news she had just received. She had been aware that major funding was flowing into *Green Force*. She had not identified the exact source of the funds but was fairly certain the money originated outside the country. Her instincts said North Korea, but she still needed definitive proof. She called Alex and briefed him on the development.

"I'm fairly certain General Jon Yong-nam is behind the whole thing," she told Alex, "so watch yourself. They're not above crashing the loop…I think they'd do it in a heartbeat, and they've killed before. And remember, Lars Watson is out there somewhere."

Mabel called up a secure information file on her link, displaying an aging General Jon Yong-nam. He was one of the close young advisors to Kim Jong-un when he assumed power back in 2011, while still in his late twenties. Although both Kim and Jon were on the downslope of middle age, Kim still ruled North Korea absolutely, and Jon, well, nobody really knew what Jon did, but he seemed to have his hand in virtually everything relating to the World outside the DPRK. Mabel's sources told her that the DPRK had long since joined the ranks of the World's elite hackers. Its increasing commercial capability to put heavy loads into orbit was a growing factor that Slingshot was directly impacting.

JARVIS ISLAND COMPOUND

Sixty-five hundred kilometers away, Alex sat at his desk in his Jarvis office, pondering his conversation with Mabel. He had tried many times in his adult life to understand the mentality that destroys in order to save. He had no trouble with the concept of "saving the planet." In fact, over the years, he had taken specific steps in his own life to be part of the solution rather than contributing to the problem. Slingshot, the culmination thus far of his life's work, was a tribute to this mindset. In fact, even the underlying basis for Slingshot was squarely within the "saving the planet" paradigm. To Alex, it was intuitively obvious that the asteroids held virtually every kind of raw material humankind could ever need, in abundances that staggered the mind. In fact, Alex had calculated that there probably was sufficient material in the asteroids to construct a Type I Dyson Sphere around the sun. *Dreams for the far future*, he thought, resisting the urge to daydream about the engineering complexities of constructing a sphere consisting of billions of individual parts entirely enclosing the sun. It was completely obvious to Alex, however, that the mineral wealth contained in the asteroids could easily supplant Earth-based mineral extraction with its consequent damage to the planet's ecosystems, if there were an economically viable way to mine and bring this vast wealth back to Earth.

That was the ultimate function of Slingshot—inexpensive, routine transport into space from the Earth's surface. Ultimately, Alex reasoned, virtually every polluting manufacturing process would be handled off-planet—on the Moon's surface, somewhere in the asteroids, on one of the moons of Jupiter or Saturn, in habitats constructed at the Earth-Moon La Grange points, or even located in deep space away from damaging gravitational stresses. The point was that the Earth could finally become a genuine Garden of Eden, a serene Central Park in a solar system of thriving commerce.

※

Alex pulled himself from his reverie and called Sanchez. "Tex, what's your status?"

"Mr. Yoon and I will depart in about thirty minutes, Boss. You gonna see us off?"

"Be there in a few, Tex," Alex responded and left to find Bruce and his father.

He found them in the library—a room with many of the amenities of libraries of old. The walls were surfaced with warm rosewood, the carpet was plush and deep, the chairs were leather and maroon. The only thing missing was leather-bound books on the shelves. In fact, shelves were missing. In this library, ambiance was what ruled. A crew member who wanted to spend a quiet hour reading could sit in the library, call up virtually any book ever written using the personal Link worn by virtually every individual in the civilized world, and read while relaxing in the warm atmosphere of an age gone by.

Bruce and his father were not reading, however. They were deep in discussion about Slingshot. Before them hung a holodisplay of the entire project, and Bruce was relating to his father some of the more interesting aspects of this gigantic project. Mr. Yoon appeared genuinely impressed, not only with the project, but with his son's comprehension of its details. They stood as Alex entered.

"Mr. Regent," Yoon said, "I cannot thank you enough for the opportunity you are giving my son." He bowed slightly, and then held out his hand.

Alex shook his hand warmly and promised to look after Bruce and keep him out of trouble. A few minutes later he stood with Bruce on the baking tarmac with towering thunderheads to the south and east as a backdrop, watching Sanchez pilot *Floater Two* into the azure western sky, carrying a proud father back to his familiar commercial world, but with a big difference that would lighten his step going forward.

SEATTLE—SMITH TOWER

A conference room occupied the north side of the building, twenty-one floors below Mabel's ornate desk, just to the east of one of two open atriums that extended all the way down to the third floor and up to the twenty-first. It was one of seven interconnected office spaces comprising the north and east sides of the fourteenth floor of Smith Tower. Two of the offices were flanked by stairwells and faced a bank of eight highly polished ornate brass elevators. Like the other doors in Smith Tower, the office doors were steel, molded and hand-finished to look like highly grained mahogany. They were the original doors installed back in 1914, as fully functional now as they were then. The plate on the left door immediately facing the elevators read simply *Environment, Inc.* The hallway ceiling was ornately carved wood with inset modern lighting fixtures, and the walls were trimmed with Alaskan marble set in paneled wainscots.

The conference room by the open well contained a large round table. Equipped with a holoprojector and surrounded by plain chairs, the table occupied the center of the conference room. A utilitarian office connected to the conference room was sparsely furnished with a simple holoprojector-equipped desk and chair, a plain cabinet, and a couch with a low occasional table filling the space in front.

Quinton Radler sat with his back to the eastern window studying a holodisplay in the air over his plain desk. He was in his mid-to-late-thirties, slender, with brown hair and a short-cropped full beard, and was dressed casually in the current fashion for Seattle business people. The display he was studying would have greatly interested Mabel twenty-one floors above. Radler was reviewing the *LLI* timetable—not the one available to anyone with a Link, but the confidential timetable that only the top echelon of *LLI* could access. He seemed particularly interested in a waypoint identified as *Ascension*.

Radler spent the next half-hour calling up background material on Slingshot. Although it seemed clear to him that Ascension meant the commencement of operations to raise Slingshot's rails to orbital height, Radler was the kind of person who checked his facts, obsessively at times, always making sure before he took action. From his first day with EI, Radler was surprised to discover that *LLI* operated almost entirely in the open. The concept of proprietary information, of keeping your hand close to your vest, seemed foreign to Mabel. Radler valued physical fitness, so much so that he rode a bike to work every day, even in the rain. He had little sympathy for people who did not pay the same attention to their physical appearance. He was contemptuous of Mabel because of her size, typically referring to her as that "fat broad." It never occurred to him that his contempt for Mabel's physical appearance might be coloring his assessment of her capabilities.

Once Radler was satisfied that Ascension meant the raising of Slingshot's rails, in effect the commencement of actual ribbon movement, he called a meeting in the conference room. Radler was entirely comfortable running the show through his Link, but he frequently employed the Harvard Business School lesson that nothing beats a face-to-face meeting to get things accomplished.

EI was structured with three legs. Its general focus was to stop *LLI* by any possible means, so long as the connection back to the DPRK remained invisible. The point of their spear was the radical environmental group, *Green Force*. Darius Gotch headed up this effort. His office was immediately next to Radler's with a connecting door and a door into an L-shaped hall. Gotch walked across the hall into the conference room. He was of medium height, slender, with very short-cropped dark hair, and looked every bit the eternal Yale "Preppie" that he was. His unofficial title was Head of Site Sabotage—Head of SS, something he was secretly proud of.

Katelynn Leete's office connected to Gotch and opened into the hall. She managed manufacturing sabotage efforts, and seemed ideally suited to the task. A slender bookish woman standing 175 cm, Leete dressed with a masculine flair that diminished her bosom. She wore her brown hair shorter than current feminine fashion dictated, and her make-up was so minimal that her colleagues rarely noticed. Leete arrived in the conference room right after Gotch.

Dane Curvin was in charge of personnel—ensuring that *EI* people were in position throughout *LLI*, always available to carry out surreptitious tasks assigned by Gotch and Leete. Like his colleagues, Curvin had a window looking east, but his southern wall was graced with a window overlooking the larger twenty-one-story open well that formed the core of the lower part of Smith Tower. His office also opened into the L-shaped hall. Curvin was Gotch's Preppie twin and fellow Yale graduate, but unlike Gotch, Curvin continued with Yale Law school. He stood 180 cm, was clean-shaven with sharp features and piercing blue eyes. He wore his brown hair cropped short like Gotch. Curvin followed Leete into the conference room, and all three took their places, awaiting Radler.

The ubiquitous Link had dramatically changed modern office environments. Offices still employed paper, but the effusive use of paper that characterized work environments of a quarter-century earlier had been almost entirely replaced by electronics. Virtually everyone in the modern world wore a Link about the size of a twentieth-century wristwatch—usually on the wrist, but it could be worn anywhere on the body or incorporated into any clothing or jewelry piece. The Link was a gateway to the Web—a vast, all-encompassing virtual reality that had evolved from the Internet and World Wide Web of the early twenty-first century. The Web contained every book ever published, and most written but not published, every piece of medical and personal data on everyone alive and most who ever lived—it was the repository for all data from every source throughout the world. The Web was controlled through massive molecular computers located in undisclosed locations on the Isle of Man, with backup facilities in America, Europe, and Asia. It was not a government operation, although Whitehall had its fingers in nearly every part of the system. The Web was run by a consortium of large firms, each of which held one seat on the board. In effect, it received its funding from the pre-tax profits of several of the world's largest companies. Rumor held that as soon as it became practical, the Web would move off-planet, sufficiently close to avoid significant transmission delays, but entirely away from any government influence. Nevertheless, the Link formed the backdrop for virtually everything in modern life.

While the three department heads waited for Radler, they called up their individual Links and continued the tasks on which they were working when Radler called the meeting. As Radler stepped through his private door, the three holodisplays folded into themselves and vanished. Radler took the empty seat and activated the table's holoprojector.

"This," he said, "is the target." He pointed to the holodisplay filling the table center. Despite his austere style, Radler was given to occasional hyperbole. "And this is our goal." The three-dimensional holographic image shifted ever so slightly, and the four conspirators watched the elevated rail tear itself apart. The 5,000-kilometer ribbon parted near the *Earhart Skyport*. The bitter end of the launch rail continued on to *Noonan Skyport*, and then whipped off to the east, pulling the entire length of the ribbon from the western sheath. It carried on in a slightly parabolic path, into cislunar space, and then into an eccentric solar orbit, leaving a scattered trail of debris in its wake. The four sat quietly, in awe of what they had just witnessed.

Radler broke the spell. "We have freed Watson," he told them. The three looked up with surprise and interest. "He's an idiot, you know," Radler continued. "And you know what they say—Stupid is forever." He looked at Gotch. "Darius, I want you to arrange a meeting with Watson. Clown it up. Give him some Hollywood spy stuff—make him b-e-l-i-e-v-e that he's playing a kingpin role to save the fucking Earth!" He turned to the holodisplay. "This thing will commence Ascension in about three months. Between now and then, I want you, Katelynn, to cause havoc with their manufacturing process. Delay Ascension from your end. Let them find your people; get them fired, and get them replaced with Dane's recruits. Keep them guessing one hundred percent of the time." He looked around the table. "You guys earn your pay now."

Radler turned to Gotch again. "I want you to set up Slingshot for failure on Ascension—what you saw in that simulation. Do it smart. Give Watson whatever he needs by way of ships and submersibles. Just make sure that he thinks it's his idea. Funnel the funds to him through the dummy support network we created the last six months." He turned to Curvin. "Dane, send your recruiters to Berkley, Stanford, Boston College, Columbia, Harvard and Yale, Evergreen—Hell! You know the list—all the universities that are producing the illiterate idiots flooding our modern society. Recruit a gaggle of pious little environmentalist pricks to man Watson's navy. Pump them full of ecobabble, give them to Watson, and point him in the right direction with specific instructions for what to do when he gets there—and make him think it's his idea, of course." He looked over at Gotch. "Yeah, I know—you and Dane coordinate the details between you."

JERSEY CITY—NEW JERSEY

The helicopter carrying Lars Watson had whisked him away from in front of the New York State Supreme Court building on Foley Square, swept low across Duane and Rockefeller Parks, and across the Hudson barely above the waves, lifted up over the Hudson Exchange, and then flew low and fast over Hamilton Park, Jones Park, across Kennedy Boulevard to the Holy Name Cemetery in Jersey City. As the helicopter landed, the pilot shoved an envelope into Watson's hands and told him to park himself somewhere inconspicuous and read the contents. Watson had no time to examine the envelope's contents before he was unceremoniously shoved from the aircraft.

He found a quiet bench on a triangle marking a meeting of three trails near the park's center. He sat and gingerly opened the envelope. The first thing he saw was a sheaf of hundred-dollar bills. A quick count came to at least $10,000. A short note was the only other thing in the envelope. The cryptic message said: *Take Whitman Ave from the NE corner of the cemetery. Turn right on Sip Ave. Go five blocks to Garrison Ave. Go left for three long blocks to the Ramada Inn of Jersey City. Give the front desk your name, pay for your room in advance for one week, go to your room, and wait for further instructions. DO NOT LEAVE THE HOTEL!*

CHAPTER FOURTEEN

JERSEY CITY—NEW JERSEY

Lars Watson had been in his room for two days without hearing from anyone. He was bewildered by the turn of events. He was deeply bothered by his ignominious capture in the South Pacific, and simultaneously thankful that Alex had actually taken care of him following his bends hit. Sure he knew that Alex had caused it in the first place, but Alex could simply have left him there to die. Watson was distraught at his mistreatment from Montague, and he harbored a palpable lust for payback. At the same time, he felt no remorse for his sexual shenanigans with the females of his clan. In fact, he filled his hours of restless waiting with wistful reveries of his sporting activities with this or that girl. Of them all, Francesca most filled his thoughts. Her little perfect body, her guileless sex play, her uninhibited release—he couldn't keep his mind off her.

During his fateful dive at Buoy 1528, Watson had left his Link topside. He never got it back. Without his Link, he felt cut off from the rest of the world, half his normal self. Things had happened so fast on Jarvis that he had not had any chance to request its return. When he asked Montague on the *Gulfstream*, all he got was a sneering laugh. During the days leading to his *Habeas Corpus* hearing, nobody would listen to him, and in the chopper ride to New Jersey, there was time for nothing at all. Without his Link, Watson was helpless to make anything happen.

Watson was also bewildered by the $12,000 (as it turned out) that someone with a chopper and obviously vast resources had given him. Watson had always been good at extracting money from widows, divorcees, and naïve kids. Because it was for a good cause—saving the Earth—he didn't care that they often could not afford their gifts. Let the State take care of them; he needed their money to further his just cause. But he had lost the *Green Avenger*, his immediate crew was gone, and he was holed up in an out-of-the-way fleabag somewhere in New Jersey, with no contacts, no help, and no Link to summon them.

The weather had turned, and rain plummeted against Watson's window. Dark rain clouds lay low overhead, blanketing the sky as far as he could see. The world was gray, not only because of the weather, but because everything was gray for Watson, all the time. He had been born totally colorblind. When Watson looked at the world, all he saw were shades of gray. It was a secret he had told no one, a secret he had learned to live with. He knew, for instance, that red lights were always at the top and green at the bottom, even though to him they appeared identical. Getting dressed could be a problem, since what looked coordinated to him sometimes turned out to be violently clashing colors. So he had learned to live with the handicap. Watson felt closed in; he missed the open air and the uncluttered ocean. But the cryptic note had been clear. *DO NOT LEAVE THE HOTEL!* Watson figured that whoever wrote that note had enough power to enforce it. He tried to diagram his situation on a sheet of hotel stationery, but he got no further than a circle representing himself, with links to Francesca, Tiffany, Carmina, and the other girls he remembered.

Watson heard a sound outside the room door. He turned to see a slip of paper that had been pushed under the door. Quickly, Watson opened the door and looked up and down the hall. Nobody, not even a cleaning person. Watson closed the door and picked up the slip. It read: *Buy a bouquet of flowers at the flower shop in the lobby. Then return to Holy Name Cemetery at the point where you were dropped. Locate a grave to the west two rows in from the road with a large bouquet of purple and blue flowers. Present yourself at that grave and pay your respects for five minutes. Leave your flowers, and retrieve a Link attached to the back of the vase holding the purple and blue flowers. DO NOT ACTIVATE THE LINK THERE! Return to your room and secure your blinds. Then activate the link and follow your further instructions.*

SEATTLE—SMITH TOWER

"Where is he now?" Curvin asked his man in Jersey City.

The image in his Link grinned and answered, "He just entered the hotel. He picked up the Link and made his way back to the hotel just as you predicted. He doubled back several times, looked at the reflection in a couple of display windows, even walked through a grocery store for a few minutes." The observer laughed. "Just like that old TV series, Get Smart." The simulacrum started laughing. "This guy's a piece of work!"

"Good job—thanks." Curvin terminated the link and leaned back in his chair, waiting. A few minutes later, his holodisplay flashed, and the upper right corner formed its own independent holodisplay. Below the display, the words *Watson Link* appeared. Curvin punched an intercom button. "Hey Katelynn, come in here and look at this."

The side door in his entryway opened, and Leete entered his office. "Whatcha got, Dane? Whatcha got that's so impressive?" She looked at his holodisplay as she approached. "What's that?" She walked around the back of his desk to see the front of the display.

"Watson's Link, Babe." He grinned at her. "How about dinner?"

"You know better than that, Dane. You got the wrong plumbing," she said playfully. Then becoming serious, she said, "That's pretty slick." Turning, she added, "The machinists at *LLI* are going on strike. I got to monitor their progress. Later." As she walked through the door, she said over her shoulder, "Change your plumbing, and I'll seriously consider it."

"Dyke," Curvin muttered under his breath. He adjusted his Link so Watson's holodisplay and his own reversed places. Watson was searching out the current status of Slingshot. Curvin activated his intercom again. "Darius, pull up…," and he gave him the crypto-co-ordinates to view Watson's Link. "He has no idea," Curvin added.

Gotch commenced actively monitoring Watson's Link. His simple plan was to feed doctored information in response to Watson's info search requests. He figured that Watson was so disconnected from the real world, and so completely reliant on the Web, that he would not even question what came over his Link. First, however, Gotch sent an encrypted message to Watson, telling him to stay in his room for another day, and then to check out and go about his normal business, but with the lowest possible profile. From time to time, Gotch told him, Watson would receive anonymous contributions that would give him the funds he would need to continue his important work to save the planet.

AMERICAN SAMOA—SOUTH PACIFIC

As Watson roamed the Web collecting information for his operation going forward, he began to build a mental image for how best to reach his goal of stopping Slingshot. Carefully guided by Gotch, Watson came to understand that his next move would be to set up a catastrophic failure of Slingshot during Ascension. As his understanding of Slingshot increased, he realized that if he could significantly weaken several of the tensioner cables that kept Slingshot from tipping over during Ascension and kept it vertical during normal operation, Slingshot would crash. Further research, unknowingly guided by Gotch, brought him to the realization that the best way to accomplish this surreptitiously was to use a deep submersible to cut partway through several tensioner cables right at the massive concrete anchors at the Western Complex. Further research convinced him that instead of cutting the cables, which was risky at best, he could attach explosive cutters that would be triggered by significant tension. Then, during Ascension, sudden tension on these cables would trigger the explosive cutters, causing the cables to part with the resultant crash of the entire rail.

Watson was highly pleased with his incipient plan, especially so when an anonymous donor gave the aging, but fully functional, working deep submersible *Alvin* to *Green Force* to be used in its ongoing deep ocean research related to deep-water pollutants that might be destroying the Earth's oceans. *Alvin* was delivered by air to Pago Pago in American Samoa, where the Philippine-flagged ship *Aku Aku*, which had been especially modified to service *Alvin*, was waiting, tied up at Pago Pago Harbor on the north shore of the bay. Normally a mothership for a deep submersible carries the little sub on its deck, and is outfitted with special cranes that can lift the sub and lower it overboard into the water. The *Aku Aku* was different. She was outfitted with a mid-ships well decked over at the top flush with the main deck. The hold just aft of the well had hydraulic doors that opened into the well, and a gantry crane designed to carry *Alvin* from the hold into the well and launch it from there. Upon *Alvin's* return, the gantry would pick up the sub and return it to the hold. With this capability, *Aku Aku* could launch *Alvin* just outside the Western Complex, if she could come up with a sufficient reason for being there in the first place.

Watson flew to American Samoa and spent two weeks under instruction in the bay learning how to operate *Alvin*. He discovered that he was actually pretty good at it, and by the end of the second week, Watson was able to descend to the bottom, locate a prepositioned marker, and successfully emplace an explosive cutter around a cable stretched to a surface buoy, without snapping the cable in the process. Getting back to *Aku Aku's* well was a bit tricky, but once Watson grasped the concept known to every instrument pilot, trust your instruments, he was able to return to the well with *Alvin's* ports blanked out by just keeping a dot on his panel display inside a lighted box. He was ready.

That was when Gotch slipped the next piece of the puzzle into the equation. Watson determined that since the tensioners were so critical to Ascension, *LLI* would very likely be paying close attention to the anchors, cables, tensioners, and everything surrounding them. This meant he would need a significant distraction that would deflect attention away from the tensioners and toward something else. As Watson pondered this problem, he found that *Green Force* was becoming the happy recipient of dozens of large boats and small ships. Through its front charities, *EI* began to purchase a significant percentage of ocean-going vessels that were for sale in marinas around the world—not so many that it brought attention on *EI*, but sufficient to cause a noticeable bump in yacht sales. Curvin recruited young radical environmentalists to man the vessels, and found moderately experienced skippers to run them. They were outfitted with sampling equipment designed to find plastic debris in the ocean. Radler had caused several scientific articles to be circulated in peer-reviewed scientific periodicals that revived the discredited early twenty-first-century theory that manmade plastic debris was entering the ocean food chain, causing significant damage. According to the articles Radler circulated, the focus of the damage seemed to be near the Equator in the Central Pacific.

Watson came to realize that he could use these hundreds of vessels converging along the location of the submerged rail as the distraction he needed. Consequently, he set about creating a coordinating team of *Green Force* members who would ensure that as many vessels as possible would arrive along the submerged rail position just as Ascension commenced. He didn't stop to ask why his luck had changed so dramatically, or how such a fortuitous environmental fleet turned up at just the right time. He was too delighted with the turn of events to question how or why.

SEATTLE—SMITH TOWER

In the Smith Tower fourteenth floor conference room, Radler received a full briefing on the status of the project. Leete reported that the Seattle strike had set *LLI* production back a full month. Failed pipe sections became such a large problem, she reported, that they had to shut down the production line to solve it. Unfortunately, she said, they replaced the workers with automation, and that solved their problem and ended that particular line of sabotage. Similarly, she explained, she had shut down their magnet production, but that, too, had been automated. In fact, she reported that she finally turned her efforts in a different direction since *LLI* seemed to have covered the manufacturing process too well to penetrate. She explained that she had turned to the supply chain, causing a series of mishaps throughout the system that significantly delayed production. She estimated that she had set back Ascension by at least six months.

"And what did it cost to do this?" Radler asked.

"About a million and a half," Leete said. "Quite a bit less than I had originally thought."

Following the combined report of Curvin and Gotch, Radler asked Gotch, "So, how much did you spend?"

"About ten million, plus the cost of *Alvin*—another three million. But," he added, "at the end, we can sell most of the assets to recoup at least sixty percent of what we spent."

"Not bad, guys," Radler said, looking at each of them. "When does it all start?"

"About thirty days," Gotch answered, "but Regent sets the pace, of course."

BAKER ISLAND—MARGO'S QUARTERS

"There are hundreds of them," Alex said, referring to the fleet making its way to the Equator between Baker and Jarvis. Margo was at her desk in her quarters on Baker Island, while Alex and Lori were comfortably ensconced in chairs facing Margo. Bruce and Carey were also present. Bruce had pulled a chair up close to Lori, dazzled by her femininity. Lori was demurely clothed on this morning; nevertheless, her aura had its effect on Alex and even on Margo. Carey sat near Alex, apparently too much the nerd to fall under Lori's spell. Dex moved about the room, recording background material, enjoying the multi-level interplay at Margo's desk. "Almost out of nowhere," Alex continued, "worldwide environmentalists have focused on oceanic plastic debris. That silliness was addressed decades ago, but here they are once more, straining at the harness to get back at it again. And why here? Why now?"

"Mr. Regent…" Carey slipped a hesitant hand into the air.

"What is it, Carey?"

"I been, like, checking this thing out. Here's what I found out so far." Carey told them that the current plastic rage originated with four articles in four peer-reviewed journals. When he followed up on the researchers and their data, he found what appeared to be real data points, but when he hacked into the underlying files, he discovered that all the points seemed to have been generated by a sophisticated algorithm. Carey made no claim for understanding the algorithm, but he stated confidently, "That's not real data. Somebody made all that up."

"Are," Alex said.

"Whaddya mean, are?" Carey asked.

"Those *are* not real data, not *is*…data is plural, not singular. Data *are*, not data *is*." Alex smiled at him. "It's a pet peeve of mine, Carey. Carry on."

Then Carey told them how he had followed up on where all the boats came from. "I came up with this list of charities," he said, projecting the list in his holodisplay. "I hacked into their files, and it turns out," he said with a triumphant grin, "that they're not real charities at all. They're just fronts for *Environment, Inc.* and," he stood up, "*EI* fronts for *Green Force*." He gave a little bow. Then he sat back down and continued, "When I tried to hack into *EI*, I came up against an impervious firewall. No way anyone can get through it. No way…" His voice trailed off apologetically. "No way…"

"Don't worry about it, Kid," Alex said. "This is fantastic information." Turning to Lori, he said, "Lori, please sit on this for the moment. You'll get first crack at it—I promise."

Bruce looked at her with adoring eyes. "I'll make sure of that," he told her.

AMERICAN SAMOA—SOUTH PACIFIC

With unseen guidance from Gotch, Watson came up with a scheme to hire out *Aku Aku* as a media vessel for bringing reporters and other media types to watch Ascension. Without arousing suspicion, he obtained permission from Alex's staff to bring the vessel to within about a kilometer of the Western Complex during Ascension. This permission went unnoticed by Alex, who was grappling with all the last-minute coordination requirements that would make Ascension happen. His biggest problem, however, remained the fleet of several hundred vessels that were closing on the submerged rail. The entire project was critically vulnerable to interference as the rail began its slow rise into the air. Almost anything could stop it at this point—a heavy line tossed over the casing, a boat ramming into the casing, a deliberate puncture before it left the water, small arms fire as it rose into the air. The list went on, and Alex was desperate to solve the problem.

Alex was about to present the problem to Mabel when Carey approached him. "I been thinkin' about the boat problem, Boss. They all got Links, you know. All of 'em. Like, every boat and every person. No one goes without 'em these days. I think I can take over the channel they're usin'—maybe not forever. Like, the Web is set up to make this very difficult, but I think I can do it, like for a while, anyway."

"And exactly what do we do if you accomplish that?" He definitely had Alex's interest.

"We tell 'em to stand off, like a klick from the buoys, and await further orders. Sumpin simple like that. Sumpin simple that sounds, you know, like it came from Lars."

"How long will it take you to hack in like that," Alex asked.

"Ten, fifteen minutes max, that's all, Boss."

"So we get everything ready," Alex said. "Get the rails ready to ascend. Then you hack in, and we send the entire fleet to the perimeter, and then we get the rails into the air and out of their reach as fast as possible."

"That's how I see it, Boss." Carey sounded inordinately proud of himself.

BAKER ISLAND—MARGO'S QUARTERS

Alex discussed the plan privately with Margo and over a secure Link with Klaus. Neither could detect a flaw in the plan, but both pointed out that it all hinged on Carey's ability to hack into the Web. "He seems pretty confident," Alex told them, "but we can't do a test run for fear that the Web will detect the hack and close that avenue."

Alex set up a meeting with himself and Margo in her quarters, with Klaus and Stewart attending holographically. He let Lori, Dex, and the two boys be present, but cautioned them to keep in the background. "This one's off the record," he told Dex, and watched as Dex shut down and removed his holocams.

"No Link recording," Alex cautioned Lori. She smiled at him warmly and pulled out a notepad and pen.

"I'll do it the old-fashioned way," she said.

When everything was set up, Alex linked to Mabel so she could participate in the meeting.

Alex convened the meeting and began to take reports from Margo and Klaus. All the anchors were in place, Margo told Alex. The tensioners on the downslopes of both complexes and the north-south tensioners every 600 klicks along the rail—they were all in place and tested. Klaus reported that both OTEC generators were up to full power and had been so for two weeks. The linear drivers were completed and had been tested as of the previous afternoon. Klaus reported they had actually moved the ribbon several meters without incident. Stewart confirmed the test from his end, and also reported that the socket for the *Jarvis Skytower* was fully in place and ready to go. Furthermore, he said, everything was in place for lifting the pieces that would become the *Noonan Skyport*. Klaus affirmed the same details for Baker and the *Earhart Skyport*. They both reported that the gigantic reels of skytower cable were ready in barges at both Baker and Jarvis. The boost and lift cable assemblies were in barges at both islands as well.

☀

Alex went down his lists guided by his Link, stopping from time to time to ask pertinent questions. Initially, Lori was able to give the session her full attention, but with the passing minutes, the detailed minutiae began to pass over her head. She allowed her gaze to sweep Alex, slowly taking in his every feature. Part of her wanted to check him out ever so much under the most intimate of venues, although she was despairing of that ever happening as time passed without the closure she craved. Then she let her gaze move to Margo. Well she remembered their intimate night together, and thinking about it now brought a flush to her face. Margo picked up on it and winked at her without distracting Alex. Lori relaxed into her reverie as the flush faded, and simply floated through the meeting, letting the words and interplay leave their impressions on her mind. She took occasional notes to give herself memory points for later recall, but otherwise, she simply moved with the flow.

☀

Bruce kept his attention glued to the proceedings, knowing full well that he would be called upon to present various aspects to the media in the coming days. He was also keenly aware of the effect Lori was having on him. He caught her examination of Alex, and was highly turned on by her exchange with Margo, but Bruce managed to keep his cool and refocused his attention on the briefing.

As Bruce listened, he was increasingly amazed at the utter complexity of what was about to happen. A flexible tube consisting of close to eight-and-a-half-million iron pipe segments, each sixty-centimeters long and five-centimeters in diameter, would be connected by slip joints, with alternating male and female segments; an evacuated sheath around this pipe with approximately 167 million special magnets holding the pipe segments suspended inside the sleeve; another evacuated sleeve around this for the atmospheric legs; a frangible outer sleeve around the orbital rail that will dissolve in the sun's ultraviolet after being exposed to upper atmosphere ozone; the entire thing driven by linear drives powered by OTEC generators—it totally boggled the mind. Bruce simply could not believe his good fortune to find himself a part of the inner circle for this project. He was so overawed, in fact, that he was able to drag his attention from Lori to focus closer on the proceedings. But not for long. The rustle of Lori's low-cut blouse split his attention once again. He snuck a sideways glance at her.

She caught his movement and stuck out her tongue at him for a brief moment.

✺

Finally, about two hours after Alex called the meeting to order, it wound down. Mabel signed off. Margo slipped into the bathroom, and when she came out, Lori entered. Margo suggested refreshments and kidded Klaus that if he hurried, he could just get there in time for a nightcap. Alex found the *Islay*, and introduced Dex, Bruce, and Carey to "the nectar of the Gods," as he called it. Margo poured two goblets of deep burgundy Merlot for herself and Lori. Alex hoisted his Glen Cairn Glass, one-third filled with liquid amber.

"*Ad astra*," he intoned solemnly. "To the stars."

PART II

"...and over the birds of the air..."

Image 9—The Earhart Skyport on the Slingshot Launch Loop

CHAPTER FIFTEEN

EQUATORIAL PACIFIC—ABOARD *AKU AKU* SOUTH OF WESTERN COMPLEX

The *Aku Aku* signaled her arrival several kilometers south of the Western Complex. Three days earlier, she had departed Pago Pago with thirty reporters and other media types. She made a beeline for the Western Complex at her cruising speed of fifteen knots. The passengers knew nothing about her secret cargo. They never met Lars Watson or any members of his support crew. Their three-day cruise was marked by good food, good wine, and good company. As they arrived off the Western Complex, more than a few *Alims* were swallowed to clear out the cobwebs, and black coffee flowed freely.

Down in the support ship's bowels, the *Green Force* crewmembers feverishly completed the launch checklist. Battery charge, air banks, emergency ballast jettison… Watson paced back and forth, impatiently waiting for the process to be finished. His lead technician exited *Alvin* through her top hatch and gestured to Watson.

"You sure you don't want me to go with you?" he asked. "It's not your practice depth of a hundred meters. You'll be five thousand meters below the surface out there." He paused, pointedly. "That's a whole new ball game."

"I can do it!" Watson snapped. "Let's get this show on the road." He grabbed his backpack and strode toward the ladder angled against *Alvin's* flank.

"I'll stay up here till we close the hatch," his lead technician said.

Watson climbed the rickety aluminum ladder, stepped onto the small platform just behind the flush hull opening, handed his pack to the technician, stepped over the lip, and climbed down the ladder into *Alvin's* titanium passenger sphere. The technician dropped his pack down and then swung the hatch closed and sealed it. The hatch into this current titanium sphere was a half-meter in diameter, slightly larger than hatches on previous versions, and like them, sealed from the outside so that when the sub was at depth, the water pressure kept the seal tight.

With only himself inside the sphere, Watson had plenty of room. He sat in the pilot's seat, donned his headset, and gave the signal to commence. In addition to several thick conical Lexan ports, he had a flat panel display that he could shift from his internal computer output to the view from any one of several outside video cameras. Since the submersible was so old, it was not provisioned with holoimaging, and its computers were virtual anachronisms, but they still did what they were designed to do, and Watson was familiar with their idiosyncrasies as a result of his crash training course in Pago Pago. He activated one of the external camera displays and sat back to await the launch.

※

Outside the sub, a technician held a heavy rubber-sleeved gantry crane remote control connected to an overhead cable run. He pushed a button and moved the trolley so the large double hook was directly over the heavy T-bar behind *Alvin's* hatch. He lowered the hook, and the lead technician still atop the sub slipped it around the T-bar. The crane controller then put a bit of tension on the line and waved the lead technician down from the sub. The crane operator signaled to two technicians on either side of the sub to take tension on their double stabilizing lines, and then he lifted *Alvin* from her cradle. Hanging on the massive double hook, *Alvin* swung gently as the *Aku*

Aku moved with the ocean's surface. The techs laid into their lines, and the swinging stopped. The crane operator signaled the bridge, where the deck officer slowly swung the ship to point into the southwestern swell. He slowed *Aku Aku* to bare steerageway. Down below, the hydraulic door to the well opened quietly on its well-lubricated rollers. The crane operator commenced moving *Alvin* out into the well. The line handlers followed the sub, maintaining tension as they moved out on catwalks hung from the overhead that left only a few centimeters of headroom. When *Alvin* was centered in the well, the crane operator lowered it into the water.

On a signal from Watson, the crane operator lowered the double hook to disengage it from the T-bar, and moved the trolley a half-meter back to clear the little sub. The line handlers retrieved their lines from the deck rings on *Alvin* by pulling them through the rings. He signaled Watson that the hook was clear, and *Alvin* slowly slipped beneath the quiet water in the well.

EQUATORIAL PACIFIC—SUBMERGED ABOARD *ALVIN*

Inside *Alvin* Watson felt a surge of excitement when the double hook released him. He could sense the independent motion of his little sub as she reacted to the gentle swell inside the well. With practiced hands, he commenced his slow descent. The bottom was nearly five klicks below, and it would take the better part of two hours before he would spot the silty bottom.

Watson drove straight down using both vertical thrusters. At fifty meters, he noticed a significant loss of light. Something flashed across his flat panel display, but it was too quick for him to identify it. He slowed his descent so that he was nearly hovering, and scanned back through his recorded images. There it was! A dolphin…it looked like he had company for the first part of his descent. He laughed with delight and continued down. What he didn't notice were the other three cetaceans circling his little craft.

At one hundred meters, Watson activated his underwater lights and scanned about the sub. The water was so clear that the beams remained invisible. As he scanned around the sub, a dolphin flashed through his beam, briefly illuminated. Then another flashed its presence. He watched fascinated as a third and then a fourth dolphin flashed through his nearly invisible light beam. The cetaceans remained with him until he passed 300 meters. He could hear their beeping through his titanium hull, even after he had passed well below 300 meters. As he continued down, the dolphin sounds grew fainter, since they were, he presumed, limited in their diving capability to about 300 meters.

After one and a half hours of descent, *Alvin* passed 4,000 meters. Watson had extinguished his lights to preserve battery power. Outside was pitch black, punctuated by occasional flashes of bioluminescence from several bizarre deep-water fish species. One came right up to the Lexan port. It was about thirty centimeters long with at least ten centimeters of jagged tooth-filled mouth. A brilliant bioluminescent ball swung from a flexible stalk extending out of the fish's head to the front of its wicked mouth—an obvious lure for unsuspecting prey. Watson couldn't know that the light had a green undertone. To him, it was just a bright light. Although the creature fascinated Watson, he remained focused on his descent and his goal.

At forty-five hundred meters, Watson slowed his descent to a near hover and activated his side-looking sonar. He slowly turned the little sub as fingers of high-frequency sound reached out into the darkness. Halfway through his turn, he received a faint echo. He stopped his turn and concentrated on the faint return. Something below him stood out like a finger from the smooth, silted plain. Using his thrusters, Watson turned *Alvin*, so it pointed in the direction of the echo. He set the little sub to a forty-five-degree down angle, and commenced a slow drive toward the echo. He activated his high-frequency locating sonar in his bow, and swept it left and right as he descended.

There! A strong return. He eased back to the left a bit and locked onto the object. He turned on his bottom lights and his high-intensity beam. And there it was—two hundred meters ahead of him, a big, obscene concrete block, flat on the bottom, with a cable snaking upward. It seemed to be crossing his position somewhere above him. Watson quickly realized that he was in a current, being pushed away from the anchor, and turned up his thruster speed to counter it. He grasped the idea that the cable was also being pushed by the current, so that he had no way of estimating which way the surface buoy lay. He knew he was to the south of the Western Complex, so it had to be to his left. The *Aku Aku* had placed itself off the first of the tensioner cable locations, nearest the complex. Watson assumed he was at the anchor for that cable.

Watson activated his external manipulators and gingerly reached into the basket at the front of the sub to grasp an explosive cutter. Then he eased the little sub up the cable for about fifty meters. He found that the current was sufficiently strong so that he couldn't hold the sub in place and attach the cutter to the cable. After five minutes or so, he grasped the cable with his right manipulator and let the current push him and the cable to its maximum extension. This took about five minutes. Then he gingerly attached the cutter with the left manipulator. Once he had figured out the technique, it was surprisingly easy. Watson checked his dive clock. He was two and a half hours into his dive. He had at least eight additional hours of bottom time, but he had to be careful not to expend too much battery power on station keeping. For the rest of the attachments, he would be sure to grab the cable with the right manipulator and conserve battery capacity.

Retaining his hold on the cable, Watson wrapped his left manipulator around the cable loosely, as if he were holding the cable in the crook of his elbow. Then he released his right manipulator. Immediately *Alvin* pulled back and took a strain on the left manipulator. Watson drove the sub back down to about ten meters off the bottom. The current seemed to be moving generally southwest. He set the sub's course for a northeasterly track, activated his thrusters, and set off to find the next anchor, about thirty kilometers due east. While he was swinging on the cable, he set in the presumed location of the next anchor while his dead reckoning computer had calculated the speed and direction of the current. Now it applied this information to suggest a specific course for Watson to follow. After about an hour and a half, Watson activated his search sonar. Nothing. Watson's stomach dropped. He had to find the second anchor! He stopped his thrusters and hovered over the bottom ten meters below him. As he focused his attention on the bottom, he noticed that it was not moving—he was stationary over the bottom. There was no current! That meant he had been bearing too far north. The anchor had to lie off his bow somewhere to the right. He swung the sub to the right, and came back up to his six knots transit speed, continuing to search with his sonar. Still nothing. He drifted slowly up while moving forward until he was a hundred meters above the seafloor. He transited an additional hour drifting up to a hundred fifty…and there it was, directly ahead of him about nine hundred meters distant.

Watson checked his dive clock—bottom time going on five hours plus; he had about four and a half left. He tilted and commenced his glide down to the anchor. He immediately noticed that as he dove, he also swung to his left, indicating he was back in the current. He flattened his track and approached the cable at a hundred meters. With now practiced skill, Watson grasped the cable with his right manipulator and shut down the thrusters. Then he gingerly attached the explosive cutter with his left. He inspected the job closely, and then wrapped his left manipulator around the cable and let go with his right—just as he had done on the first one. Slowly he lowered himself toward the anchor.

Suddenly and without warning, as he was just a few meters above the anchor, he heard a loud crack through the titanium hull. Instantly, *Alvin* tumbled to the left, rolling completely over, once, and then again, smashing with a sickening crunch on its side into the flat concrete top of the anchor. As the sub rolled over, Watson tumbled and hit his head sharply against the hull, splitting his skin against a protruding cable fixture. When the sub came to rest, he lay dazed and disoriented, blood streaming from his scalp laceration. The lights were still on. The thrusters were still running, but his flat panel display appeared broken. Gingerly, Watson rolled to his knees and shut down the thrusters while wiping the blood from his eyes. He located the first-aid kit, retrieved some gauze sponges, and tried to stem the blood flow from his cut. After about five minutes, he had managed to stop the flow with a makeshift pressure dressing consisting of several gauze sponges held in place with a strip of gauze wrapped several times over the cut and under his chin and then around his head like a headband. It would have to do, he thought, as he looked at his dive clock. Four and a half hours remaining.

Carefully, Watson tested his thrusters. All six still worked, but no matter what he did, he couldn't budge the sub—not even a centimeter, nothing at all. Somehow, he figured, the explosive cutter had activated, and he was now trapped, helplessly wrapped in Kevlar cable five thousand meters from the surface. And to make matters worse, there was a bunch of very pissed-off people topside.

Watson looked at his dive clock—four hours left.

BAKER ISLAND—OPERATIONS COMPOUND

Alex and Carey had their hands full. They were sitting in the Baker Operations Compound, surrounded by people and equipment. Carey had been monitoring the *Green Force* fleet for the past week. Coming mainly from the east, they were now stretched along the entire 1,800-kilometer length of the double rail. He had quietly monitored their coordinating efforts, and noted that the main coordination had come from a mobile link somewhere to their southwest. Early that morning, the coordination ceased, but he was not sure whether it had stopped for some specific reason or because the vessels were in position to cause their mischief.

Alex had satisfied himself that Carey was sufficiently reliable to give him the responsibility for monitoring the fleet without his constant supervision, but he still showed up every few minutes to look over his shoulder and ask pertinent questions. Carey had just told him about the disappearance of the coordinating signals when he received a call from Margo in the Western Complex.

"Alex, the strangest thing has just happened out here. You know the *Aku Aku*, the media vessel you granted permission for observing from the south? Well, it showed up and took up station about three klicks out. Then it moved in by about a klick, showed us its stern, and I could swear it went into station keeping." Margo transmitted a holoimage of that event. "What do you think?" she asked.

"Hard to tell," Alex answered. "You've got a mild swell. Maybe the landnicks aboard her were getting seasick. Looks to me like they turned into the swell. What do you think?"

"That's not all, Alex," Margo continued. "About twenty minutes later, George and his pals surfaced just aft of *Aku Aku* and started making a god-awful racket. They were up on their tails, doing back-flips, going through every trick they know."

"I bet the press liked that."

"I know they like to show off, but this was different, Alex. A few minutes later, they all showed up right next to me. Once again, they went through their paces." Margo showed Alex a holoimage of their performance. "I think George was trying to tell me something, Alex. Look at him. Don't you think so?"

"You know him a lot better than I do, Margo," Alex said. "Look, I really don't have time for this right now…we're pretty busy over here. You know I have total faith and trust in you. Just stay on top of things out there and handle George as you think best." He paused. "Send Bruce to the *Aku Aku*. Have him nose around a bit, and let me know if something goes down." He winked at her and signed off.

Alex turned to Carey. "Carey, it's time to issue the fleet new orders. You're sure you can duplicate Watson's coordinating signal?"

"Absolutely, Boss, but look." He called up a graphic display. "I think the signal was, like, emanating from the *Aku Aku*," he said.

Alex examined the graphic, which showed lines of position for the past week. During the past three days, the lines shifted their direction, with the most immediate one pointing directly to the south of the Western Complex. "Great job," he said, clapping Carey on the back. "I think you nailed it! Set up your signal to come from the Western Complex, and lock the channel so they are unable to communicate out. Tell them you've done this so as not to arouse suspicions over here about incoming traffic from the fleet. Move them away from the buoy line five klicks."

"Gotcha!"

Alex turned away. "You ready, Klaus?" he asked without preamble.

"When you are."

"You ready, Stu?"

"Likewise."

"Margo?"

"Go!"

Alex looked around the room. "Lori," he said, "bring Dex over here. We're about to start." Alex watched her cross the room, all business and somehow all sex at the same time. He noticed the pink tip of her tongue for just a moment before she winked at him and set up for the holocast. *Why are you keeping her at arm's length?* He asked himself for the nth time.

Alex did not tell Lori about his orders to the *Green Force* fleet, but she could not help but notice that the fleet was pulling away from the buoy line. She indicated for Dex to record their actions. Alex initiated an all-circuit call—automatically overriding every link in the Slingshot complex. "This is Alex…Commence Ascension!"

※

Klaus depressed a button on his master console in the Western Complex operations room. Simultaneously, power surged from the western OTEC into the two western linear drives, and from the eastern OTEC into the two eastern linear drivers. The power surge was timed so that each motor commenced pulling the ribbon at exactly the same moment—right down to the microsecond. Very slowly, imperceptibly, the tubular ribbon began to move. Sensors along the entire five thousand kilometer rail measured thousands of times per second, and faithfully transmitted their measurements to the computer Klaus was monitoring.

Lori's eyes and Dex's holocams were glued to the wall-size holodisplay that filled one side of the Operations Center. At the moment, it showed a close-up cutaway of a linear driver, with the tubular ribbon starting to show significant speed. The sixty-centimeter segments were beginning to whiz by.

"System check," Alex said on all-call.

Throughout the system, technicians with specific monitoring responsibilities verified that their area was operating nominally, and so signaled by link. The main holodisplay shifted to show the entire project, and green points began to light up all over the project. Within about a minute, Klaus announced, "Green board!" A ripple of excitement passed through the Operations Center.

Alex shifted the display to the socket near the middle of Baker Island. It looked like a flat round concrete apron about a hundred meters across with a massive ring at the center, surrounded by ten evenly spaced smaller rings.

"When the rail rises out of the water," Alex explained to Lori, "a chopper currently waiting on this barge," the view lifted into the air with a bird's view of the island with the barge alongside the wharf in Meyerton Harbor, "will ferry the bottom end of the skytower cable to the socket, where the Baker crew will attach it to the big ring." The scene shifted to a bird's eye view over Jarvis. "The same thing here," Alex said, pointing to a barge alongside the wharf in Millersville Harbor.

"As the rail continues to rise," Alex said, "the two downslopes will begin to lift as we push the deflectors away from the islands with these ocean-going tugs." Alex indicated a large tug standing just to the west off the Eastern Complex, nosed against a large vertical plate in the water. "At that point, the tensioners, which are thirty kilometers apart on each side, will begin actively stabilizing the downslopes." Alex smiled at Lori. "This will take a while. At some point, you and Dex can go out and watch this directly."

Alex continued to walk Lori through the operation, while monitoring the entire system as the ribbon speed continued to increase.

Four and a half hours into the operation, Margo announced on all-call, "The rail to buoy tension is zero along the entire rail." Shortly after that, she announced, "The tensioners are beginning to feed cable—the rail is rising. Initiating float implosions."

Alex zoomed the holodisplay to show a section of rail near Baker. The floats were clearly visible, one every kilometer. Suddenly, all the floats in the display simply disappeared. One moment they were there, the next, they were gone.

Lori gasped. Alex grinned at her. "It's okay, Lori," he said. "That's what we expected."

Then a thin line appeared on the surface. The view shifted to a horizontal long shot. Slowly and elegantly along the entire rail of nearly two thousand kilometers, the sheath lifted out of the water. Within just a few minutes, it hovered about a hundred meters in the air, just visible to the eye if you knew where to look. The choppers departed their respective barges and delivered skytower cables that crews quickly attached to the large rings. As the downslope portion of the rail lifted out of the water, Alex glanced at his watch. They were five hours into the operation. Things were going pretty well. The rail was at about nine hundred meters, and every tensioner seemed to be operating properly.

Suddenly, without warning, an alarm sounded, and the control system announced on all-call and locally over loudspeakers: "Down tensioner three west has parted. Ascension suspended and holding."

"What happened?" Alex asked Klaus and Margo on a private link.

"No idea yet," Margo answered. "We're investigating."

The control system interrupted: "Down tensioner one west has parted."

"We're stabilizing," Margo said quickly. "We slacked two and took extra tension on rail one." She paused, and then added, "W-1 and W-3 are hanging free. They definitely parted. The automatic tensioning controller has completely stabilized the system. We need to hold here while we investigate." She paused again. "You know we have to replace the tensioners. I'll keep in touch."

"Boss, it's Bruce."

"Not now, Bruce. We're running a major problem."

"It's important, Boss, really important!"

"Okay, Bruce, make it short."

"I ran into one of my old *Green Force* buddies here on *Aku Aku*. He doesn't know I'm with you now. He showed me a minisub launch facility that can surreptitiously launch and recover a minisub. He said Lars is out there right now—been on the bottom for about three hours. Said he intended to attach tension-triggered explosive cutters to the tensioners."

"Thanks, Pal. When does he return?"

"He's supposed to do four tensioners and come back in about four more hours."

"He already got two, Bruce. Mingle with the press over there, and we'll get you back in a bit."

"It's Watson, Margo. Apparently, he still has a couple explosive cutters."

"He can't do that," Margo said, with a bit of fear creeping into her voice. "That'll crash the loop." She paused and looked directly into the holocams. "You had better bring it back down."

"Can't, Margo. The fleet will destroy it if it comes within their range again." Alex put a bit of urgency into his voice. "Margo, the cutters are tension-triggered. Ease the tension on W-5 and W-7, and then get your butt to the bottom to see what's going on!"

CHAPTER SIXTEEN

EQUATORIAL PACIFIC—SUBMERGED ABOARD *WAMPUS*

Margo kept two deep-submersible subs on the project, one at the Eastern Complex, and one at the Western Complex, each with its own eighteen-meter tender—*RV Freddie N* and *RV Amelia E* at the Eastern and Western Complexes, respectively. The vessels were virtually identical, carrying a minisub in a cradle on the deck, with an A-frame crane to hoist the sub over the fantail and into the water. Sitting on station near tensioner W-1 with the birds wheeling overhead in an azure sky, Margo anxiously awaited completion of preparations to deploy the sub. Dive Supervisor Pearl Wells was running the A-frame. She was a tall, black, tough-as-they-come Gullah-Geechee oil rig diver from the Carolina barrier islands. Pearl was an early hire for the project, and she turned out to be a superb supervisor. She wore her kinky black hair cropped close to her skull, and on the job, she would throw her shoulders into a task right alongside her men, asking and taking no quarter. She was a fish in the water, and played the sub crane like a concert piano.

The minisub Pearl manipulated over the fantail and into the water was a state-of-the-art piece of underwater engineering. Capable of diving to any ocean depth, the *Wampus* (and her sister sub, the *Wallus*) consisted of a newly developed pre-stressed transparent sphere large enough for three people, with artificial gills, thrusters, manipulators, holocams, and all the other things necessary to work at depth while sustaining life. *Wampus* was capable of ten knots, although this drained her batteries fairly quickly. She used a negative-buoyancy-assisted descent that gave her up to a twenty-knot descent rate. In oxygen-rich water, bottom time was limited only by battery charge and passenger endurance, because *Wampus* extracted oxygen from the surrounding water, and used a newly developed electronic-molecular trick to scrub carbon dioxide.

Twenty minutes after identifying the need to conduct an onsite investigation, Margo sealed the hatch. Emmett Bihm, her lead diver at the Western Complex, was supervising from the *Amelia E's* fantail, his shaved pate glistening in the hot sun. Domingo Solak, Abel Kilker, and Apryl Searson were in the water around *Wampus*, accompanied by George and the other three dolphins.

"Ready here, Bimmy," Margo told Bihm over EFCom. He gave her a thumbs-up, tossing her a Midwestern grin.

"Solak, Abe, Apryl—stay with her to fifty meters."

Margo checked her systems one final time and then announced, "Diving!"

Initially, *Wampus* just appeared to settle deeper into the water. In less than a minute, she was completely submerged, and Margo halted the descent at three meters for an external visual check by the divers.

George had made dozens of dives with Margo and *Wampus* and loved her seeming unassisted presence in the water column, since the transparent sphere all but disappeared visually underwater, although it still gave a return signal to George's high-frequency sonar. He swam up against the sphere and rubbed his belly along the smooth surface, mewing happily. As the three divers carefully inspected the various fittings and potential leak spots, George playfully nudged them if they failed to rub his dome when he passed. The divers took a full five minutes for their inspection.

Apryl swam to the front of the sphere and presented herself in as coquettish a manner as possible, what with skins and gills, and said,

"You're good to go, Margo," her otherwise cute girly voice distorted by her equipment.

"Okay, Apryl," Margo answered. She was amused by Apryl's unabashed joy of living and guileless sexuality that always seemed to surface in the face of any kind of danger. She knew the men delighted in that aspect of her personality, and suspected that several of the women did as well. As the boss, she stayed above it, but it was fun to observe the interactions.

With practiced hands, Margo initiated her dive. She kept her descent speed slow for fifty meters so the divers could keep up with her. At fifty meters, the divers gave her a quick once over, and Apryl again presented herself, saying, "See you when we see you," and crossed her arms across her breasts. Margo smiled, imagining what was likely to happen when the divers showered down a bit later, and then she initiated her rapid descent and dropped like a rock. George and his companions kept up with her for the next 250 meters, and then they shot to the surface as she continued her headlong plunge to the bottom over four-and-a-half kilometers below.

Time to bottom at her descent rate was about eight minutes. *Wampus* was designed to slow her descent automatically at two hundred meters from the bottom. In an abundance of caution, Margo always kept close watch on her descent, with her hands on the manual controls to override the automatics in case something malfunctioned. It never had, at least not yet, but Margo was determined to be ready when it did. *Wampus* slowed with noticeable deceleration and commenced a slow powered descent. Margo took over, went to hover, and activated her scanning sonars.

The holodisplay was engineered to superimpose its imaging on the transparent sphere so that the images appeared outside, in the water, as if she were looking through a telescope at actual objects. When the software could not resolve exactly what an object was, it presented a geometric shape that most nearly represented what it saw.

Klaus, at his topside console, could generate the same view as Margo. "I don't see anything yet," Klaus said over EFCom, with a clarity that made it seem as if he were inside the sphere.

"Me either," Margo said. "I can see faint returns from the more distant anchors, but nothing from down one and three. I'm going to stay at this depth and move on a southeast vector—see if I can pick them up directly."

"I agree," Klaus said needlessly, since his role was to step in should something go wrong on the bottom.

Margo kept her forward speed at five knots while saturating the water in front of her with high-frequency sound. She kept her lights off to keep from washing out any faint echoes in the holodisplay. Although she had little time to look around her, she never tired of the feeling that overcame her at times like this. She seemed to be floating in a dark void, empty except for the sonar-illuminated seafloor that appeared as a faint blue carpet beneath her, the occasional flash of a sonar-imaged deep ocean fish, and the more rare bioluminescent flash from actual fish sufficiently close to see with the naked eye. It was a world like no other, simultaneously a cocoon and a world with limitless horizons.

All her life, Margo had been fascinated by Amelia Earhart's mysterious disappearance so long ago. Every chance she got, Margo would scan another sector of ocean bottom off the west coast of Baker Island. She was convinced that Earhart's *Electra* rested somewhere on the bottom in an arc a couple of hundred kilometers or so west of Howland and Baker. So, while she searched for the anchor, part of her mind kept a lookout for the *Electra* that she was certain had to be out there somewhere. "Where are you, Amelia?" she whispered quietly, "Where are you?"

In the background, a subroutine in her Link kept track of the sectors she scanned every time she was submerged on one of the mini-subs, slowly filling in the wedge centered on Howland Island. Her mind wandered back to that fateful evening in 1937…

Focus, Girl! she admonished herself as a faint image began to form in her holodisplay.

"There! Do you see it, Klaus?" The least bit of excitement crept into Margo's voice. "Slightly to port. I see the anchor base forming—five hundred meters out." All thoughts of the *Electra* vanished as she concentrated on the forming image.

"Not yet…too bright here,": Klaus said. "No, wait…now I see it! Yep…I think you found it."

Margo sent more power to her thrusters and nosed down to intersect the anchor base. About three minutes later, she was hovering above the concrete block, looking down at a rat's nest of tangled Kevlar cable. She nosed down, looking for the cable end. She found it a few minutes later, wedged under a coiled section of cable. She picked it up with her left manipulator and examined it closely with the holocams.

"Definitely sliced," she said.

"I agree," Klaus said. "Notice how the cut face is smooth? If you had cut it with a knife, about halfway through, it would have parted under tension. The first half would be smooth, the rest ripped apart."

"I see that," Margo said.

"Yeah, this was an explosive cutter."

"Look at this, Klaus." Margo had removed some of the tangled cable. "It looks like the cable broke the beacon—it's physically smashed."

"When cut," Klaus said, "the cable must have zipped down like the end of a whip. Would have been kinda exciting if you had been nearby."

"I'm going to find W-3," Margo said. "Your guys can haul up the cable end, repair it, and install an anchor clamp and a beacon. When it's ready, dunk it. I'll locate it, pull it down to the anchor, and attach it."

Margo traversed the same route that Watson had taken earlier that day, heading generally west, but compensating for the southwesterly current. Her superior speed and state-of-the-art sonars brought her within imaging range within forty-five minutes. During the transit, her Link continued to scan for any vestige of the *Electra*, without success, except to fill in another segment of the wedge.

"Klaus, you're not going to believe this," she announced. "We've got a situation down here."

Margo activated her lights and flooded the area with brilliance. On a jumbled mess of cable atop the concrete anchor surface below her, wrapped twice around with cable, the minisub *Alvin* lay on its right side, its famous name emblazoned just below its flush hatch. "Are you getting this, Klaus?" Margo asked as she carefully inspected the trapped sub.

Margo examined the trapped minisub from all angles. She could see no light through the wedge-shaped ports, and they were too small to see inside. She brought one of her lights right up to the wide face of one port and brought a holocam up against another. This gave her a view much like looking at the world through a toot-ta-doo. Something was moving inside, but her view was too narrow to tell what. Then the inside of the sub lit up, and a face appeared in the line of vision of her holocam—it was Watson.

"Klaus, you getting this?"

"That sonofabitch!" Klaus' voice was filled with bitter anger. "That goddamn sonofabitch!"

"I'm going to cut him loose, Klaus. If he has motive power left, he can make his way to the surface. I still must check W-5 and W-7, and we need to get the tensioners reattached."

"I'm on it, Margo."

Margo moved *Wampus* back from *Alvin* and examined the Kevlar wraps. Then she reached out with her right manipulator cutting tool, and…snip, cut through one of the five-centimeter cables. Snip…she cut another strand…and snip…the last strand.

With the loosening of the tightly wrapped cable, *Alvin* rolled to its normal upright position. The hull fairing was damaged, and it looked like the port vertical thruster was smashed. Margo watched as the three rear thrusters activated, and then, without further notice, *Alvin* slipped out of sight and headed toward the surface.

"He's on his way up, Klaus. It'll take him an hour or so." Margo started up *Wampus*, tilted the little vessel upward, and ascended to two hundred meters off the bottom. She commenced a scan for W-5. "Keep an eye out for Lars, Klaus."

EQUATORIAL PACIFIC—SUBMERGED ABOARD *ALVIN*

Lars Watson could not believe his good luck. He had given up hope of rescue, and had resigned himself to a slow death from hypothermia and carbon dioxide poisoning. When the bright light illuminated the sphere's interior, he had been slumbering, not expecting to waken. He looked through first one port, but it seemed blocked by something. Then he looked through the third, and was startled to see what appeared to be a woman floating free in the water, hovering above a sled-like contraption. But that was impossible, so Watson surmised that it had to be some kind of transparent sphere. He watched the strange sub move in, and heard it cut his restraining cables. The moment he was free, *Alvin* righted herself, and Watson immediately set a southwestern course and headed away and toward the surface.

Watson had no way of knowing that Margo was fully occupied with retrieving the cut cables and reinstalling them on their respective anchors, but he took no chances. He dropped his excess ballast and drove upward as fast as he could. Under the best of circumstances, Watson had an hour's trip to the surface. Limited as he was to only one vertical thruster, even having dropped his excess ballast, Watson was facing at least another half-hour, perhaps more. *Alvin* carried an antiquated underwater telephone system that worked much like AM radio, except the carrier was a sound signal in the water at 812 kHz, where the intelligence was carried on the lower sideband. *Alvin's* previous owners had never gotten around to replacing this antique with a modern system, probably because it worked sufficiently well for their purposes, and the available funding was better used for other things. This turned out to be an advantage for Watson. Since the Slingshot people had no compatible receiver, Watson was free to communicate with *Aku Aku* when it was actually possible, since the underwater phone was severely range-limited.

As Watson passed 300 meters, he caught a glimpse of a dolphin in his beam, and when he reached fifty meters, he could hear the dolphins actually thumping against his fiberglass hull fairing. He came to a hover, which required down thrust because of the missing ballast. Carefully, Watson scanned around him, looking for the homing beacon at *Aku Aku's* well opening. He picked up a faint signal to the southwest and commenced traversing in that direction while maintaining his fifty-meter depth.

"*Aku Aku*, it's Watson," he intoned on the underwater phone. Through his headset, he could hear the strange, hollow reverberations as his signal propagated in all directions around him. "*Aku Aku*, it's Watson…*Aku Aku*, it's Watson."

He kept it up as he continued his southwestern transect. When he figured he was still about a half kilometer from *Aku Aku*, he heard a faint, rushing signal that sounded as if it had traveled the entire distance through a garden hose. "This is *Aku Aku*. We read you!"

Forty-five minutes later, the gantry crane hooks slipped around *Alvin's* T-bar and lifted the little sub clear of the water. The line handlers noted with delight that as *Alvin* left the water, several dolphins came to the surface for air, and kept their heads above water, watching their progress. Steadied by the line handlers on the catwalks, *Alvin* moved through the massive hydraulic door into the hold and was lowered into her cradle. As the hydraulic door closed, the dolphins dived. Watson popped the hatch, pulled his ascetic body halfway out of the hatch, and collapsed. His lead technician rushed to his aid, slipped a harness around him, and lowered him to the steel deck. Watson opened his eyes and whispered, "Get the ship underway… Pago Pago."

"But the media…," his technician tried to explain.

"Screw them…Get the hell outa here! Now!" And Watson drifted back into semiconsciousness.

EQUATORIAL PACIFIC—ABOARD *AKU AKU*

Two hours later, Bruce was leaning over the starboard railing about a third of the way up from the stern, chatting with Mary Martain, a cute reporter from *Me Too!*, an obscure magazine for twenty-somethings, explaining to her the intricacies of the Launch Loop. Bruce had already told Alex that they were underway, and that he had learned from his *Green Force* contact that Watson was back onboard.

"Stay with them, and see what you can find," Alex told him. "Tex will pick you up in Pago Pago."

The overhead sky was clear, but storm clouds brooded on the southern horizon ahead of them, and towering white thunderheads dotted the western horizon. Dozens of sooty terns, masked boobies, and lesser frigatebirds filled the air over the ship's wake, skreeghing, fishing, swooping, the frigatebirds always on the lookout for a loosely-held catch to steal. Below them, George and his friends cavorted in the ship's wide wake, dashing up the starboard side, squealing repeatedly. "Is that you, George?" Bruce shouted. George squeaked back and lifted his body halfway out of the water to emphasize his recognition. Martain squealed back with unabashed delight.

Bruce and the reporter continued to watch the dolphins' antics, while Bruce explained to her that these creatures had been trained by Margo Jackson, who was in charge of all the underwater stuff at Slingshot. Since she seemed so interested, Bruce also explained how

SSRS worked and how it interfaced with EFCom. Martain asked to see his SSRS unit. Beginning to think he was making progress along a non-business path, Bruce removed his bone conduction bud from behind his ear and handed the reporter the bud and SSRS unit.

At that moment, Watson came striding across the deck. "Bruce Yoon, you little traitor sonofabitch! What the fuck you doing here?" Watson was shouting. "You damn near cost me everything, you fucking little spy!" In one quick motion, Watson grabbed Bruce, slid him up the railing, and tossed him over the side.

※

Martain was shocked beyond belief at what she had just witnessed. Watson continued to lean over the railing, shouting obscenities at Bruce. She slipped inside the nearest door to get away from the crazy man and held the SSRS to her mouth. "Hello…hello!" Then she remembered that Bruce had removed the bud from behind his ear. She put it behind her ear. "Hello…Hello..is anybody listening?"

"This is Alex Regent. Who are you?"

Haltingly, Martain told Alex what had just happened. "What do I do? What do I do?" she wailed.

"Go outside to the railing and tell me what you see," Alex told her. She followed his instructions.

"The dolphin George is right next to Bruce. Wait, Bruce grabbed George's dorsal fin. Oh my God! George is pulling Bruce away from the ship. One of the dolphins is with them. The other two took off really fast towards you. Oh my God!"

"It's Mary, right?"

"Yes."

"Okay, Mary. I want you to make yourself as inconspicuous as possible. Do you have a friend onboard?" She did. "Okay, then…lock yourself in his stateroom and don't come out. Be seasick or something, but don't unlock the door for anyone but him. You're going to lose contact on SSRS. You have your Link, right?" She said she did, and Alex gave her a special Link code that would put her in direct contact with him. "Keep your Link open," he said. "I'll have Tex Sanchez meet you in Pago Pago when you arrive." He sent her an image of Tex. "Remember, stay in your friend's stateroom."

EQUATORIAL PACIFIC—OVERBOARD SOUTH OF WESTERN COMPLEX

Bruce hit the water in total shock. He knew enough to fear the propellers that would pass him shortly, and began to swim away from the ship, applying everything he had. He wasn't a very strong swimmer, and he felt terror rise in his throat. He felt a strong nudge against his stomach, and then a dolphin surfaced next to him. "Is that you, George?" Bruce gasped.

George chattered and came close alongside him. As Bruce had seen the divers do, he reached out and grasped the dolphin's dorsal fin. It was firm and smooth in his hand and easy to hold on to. George immediately accelerated away from the vessel, pulling Bruce with him, even as Bruce felt the undercurrent pull of the powerful propellers. Bruce felt a touch on his left side. It was another dolphin. He didn't know its name, but it was clear that it wanted him to grasp its dorsal fin as well. Bruce complied. The dolphins chattered to one another and squealed with a piercing sound. Bruce was sure they were communicating with their two pals. Within moments he found himself moving through the water so fast that he had to turn his head to keep water from rushing up his nose.

"Whoa, guys," he sputtered, while letting go of the dorsal fins. "We gotta figure this out." Bruce rolled to his back and grasped both fins again. "Let's try it this way," he said, and the dolphins got underway once more. Bruce found that it worked much better on his back, and both he and the two dolphins settled into an easy skimming over the ocean's surface toward the Western Complex, still some thirty kilometers distant. Several hundred meters ahead of them, and getting further ahead by the second, the other two dolphins surged forward at maximum speed, soaring out of the water every few tens of meters, side by side in a graceful ballet.

EQUATORIAL PACIFIC—ABOARD *SKIMMER ONE*

Tto get *Skimmer One* underway after leaving Bihm in charge of *Amelia E's* fantail. She knew George was with Bruce, but, in her own words, "I don't trust no damn fish—know what I mean?" Actually, Pearl and George got along very well, but she was worried for the little cub reporter, the kid with the brass ones who had infiltrated *Green Force*. He was her kind of person, even if he was a scrawny little puke. Pearl shouted with glee as *Skimmer One* lifted on cushion as she pushed the craft to its limits.

She passed two speeding dolphins heading toward the Western Complex three minutes later. *They have to be part of George's pod*, she thought, but didn't slow to check. Four minutes after that, Pearl spotted a surface commotion and pulled her throttles all the way back. *Skimmer One* settled into the water and glided up to George and his dolphin pals, with Bruce clutching their dorsals.

"Hey, Bruce!" Pearl yelled, wiping tears of joy from her glistening black cheeks. "Enjoying your swim?"

CHAPTER SEVENTEEN

EQUATORIAL PACIFIC—SUBMERGED ABOARD *WAMPUS*

Margo was fairly certain that Watson had only emplaced two explosive cutters. She reasoned that he did W-1 first, fol-lowed by W-3, and while he still was at W-3, Slingshot pulled enough tension to trigger both cutters. No other tensioners were cut, even though they all had pretty much the same tension—especially W-2 and W-4. Nevertheless, she needed to see W-5 at a minimum to be sure. Margo scanned at two frequencies, one set for the anchor beacons, including W-5, and the other set to the beacons Klaus attached to the repaired ends of W-1 and W-3. Since the beacons on the cut tensioners would be behind her, she concentrated on looking for W-5. Simultaneously, she also conducted an active high-frequency search in case the beacon had been damaged like at W-3. Margo conducted her search without sonic enhancement. In her experience, the electronics were so much more sensitive than the human ear that there was nothing to gain by listening to the rush of her search sonar, and it interfered with her enjoyment of the utter solitude she experienced in-side her transparent sphere here in the abyssal deep. And as always, her Link continued to fill in the wedge, identifying more areas where Earhart did not end her ill-fated journey.

After forty-five minutes of steady searching as she moved westward, Margo picked up the faintest flicker of a signal. She concentrated her listening beam on that vector, and there it was, broadcasting its identity—W-5. Margo set the broadcast point into her autopilot, but just before she commenced her approach, she broadened her listening beam and turned up the intensity of her active high-frequency search. If you were to ask her why she did this, she would respond that she was making sure there were no unexpected obstructions, or anything else in her path. While this would have been true, Margo really had another underlying reason.

Margo was not the kind of person to let personal matters interfere with the job at hand, but there was nothing wrong with expanding her sonar search parameters so she could eliminate another piece of ocean bottom on the comprehensive chart she maintained on her Link. So her sonar conducted the enhanced search while she drove up to W-5 for a visual inspection of the bottom two hundred meters of cable.

She dropped to the anchor, and then rapidly rose up the cable, concentrating her lights on the five-centimeter Kevlar hawser, scanning it visually as she rose. At two hundred meters above the anchor, she called Klaus. "W-5 is clean, Klaus. I think we're okay. Apparently, Watson got caught doing W-3. I'm off to find your beacons."

As *Wampus* turned, Margo noticed a faint echo about 500 meters to the southwest—at the extreme range of her high-frequency search sonar. Although it was a bit out of her way, Margo pointed *Wampus* at the echo and cranked up her thrusters. At her elevated speed, she crossed the distance in about two minutes, but even before she got there, on her display, she could clearly see the distinct outline of an aircraft. Her heart began to beat faster. Other planes had gone down here during the war, but not that many. As she drew closer, her computer began to reconstruct the image on the bottom. There it was—an *Electra*, THE *Electra*—its right wing severely damaged, windscreen smashed, vertical stabilizer missing, but otherwise relatively intact, resting right side up, almost as if it were preparing to take off. It was an incredible sight, this world-famous aircraft nearly intact on the ocean bottom, 5,000 meters below the surface, just waiting to be found…

"Klaus, Klaus! I found it! I found it!" Margo could barely contain her excitement. "Amelia's *Electra*, it's here on the bottom. I found it!"

"I can see, Margo. Words fail me. What do I say to the girl who just became world-famous?"

This was literally a dream come true. Margo had spent countless hours examining charts, calculating vectors, studying old weather patterns, reading reports of search groups, even as late as the early 2000s. There were so many theories, so many opinions.

Margo spent another five minutes examining and recording the *Electra*, and then she attached a beacon to the wreck, and vectored *Wampus* up and to the northeast, searching for W-3 beacon.

※

It didn't take very long for Margo to locate the W-3 beacon. The current had pushed it further to the northeast, but she picked it up on her search sonar within ten minutes, and fifteen minutes later, she had a firm grip on the double clasp at the cable end.

"Klaus," Margo transmitted shortly after grasping the cable end, "I think we have a significant problem down here."

"I can guess, Margo, but tell me about it anyway."

"You can see that I have gripped the double clasps with both manipulators. What you cannot readily see is that I am in full reverse, but I'm not going anywhere." Klaus chuckled. "I know you can monitor my gauges, Klaus, but that's not the same as being down here doing full reverse and not budging."

Klaus sighed. "I guess we didn't figure the current for being that strong." He paused. "Let's see…you've got a five-centimeter cross-section, and there's at least a klick of cable in the water. That's…"

"Fifty square meters of surface," Margo interjected.

"Let's assume a five-knot current," Klaus continued. "Let's call that two-and-a-half meters per second…" He did some quick calculations. "That comes to about three-quarters of a million kilograms of pressure on the cable." He laughed. "*Wampus* is a great little sub, but no way she can generate that kind of thrust!"

"I guess we should have thought this through a bit more, don't you think, Klaus?" Margo was kicking herself for not realizing what they were up against. "We wait for just the right current, or we drop another two anchors."

"No brainer," Klaus said. "Might as well get your butt back here. It's gonna take several hours."

EQUATORIAL PACIFIC—AT W-1 AND W-3

While Margo brought *Wampus* back to *Amelia E*, where Bihm hoisted her onboard, Klaus briefed Alex and got two barges underway. Since the current was not something they could fight, Klaus opted to use it instead to keep things stabilized while he conducted repairs. W-1 and W-3, although both with one free end, had stabilized in the current's tug. The rail was at about one-kilometer altitude. The two cables hung straight down for over a half-kilometer, and then curved to the north to enter the water at about a ten-degree angle some seven kilometers out. The massive reverse-locking payout reels with their supply of tapered Kevlar cable were both located several tens of meters below the ocean entry-points of the cables. Below the payout reels, the cables formed a kilometer-long current-driven catenary. The bitter ends of the cables with their double hooks were about three-quarters of a kilometer below the surface.

About one and half hours later, two station-keeping barges drove up against each cable from the inside, so the cable passed over the barge. The overhead sky was more than half-filled with fleecy clouds. Terns, boobies, and frigatebirds had followed the barges to their respective locations, where they performed their skydance that was part hunting and part just the joy of flying. In the water below, George and his pals cavorted with exaggerated excitement, knowing that something cool was about to happen. The trained dolphins had participated in the dropping of every anchor on the Western Complex, never failing to follow the anchor down as far as they could. The men on both barges were so accustomed to the birds' aerial spectacles that they hardly noticed the ballet happening right over their heads. On the other hand, they cheered the dolphins on, calling to George and the others by name, encouraging their play.

The barge on W-1 held its position while the second barge went into action. The crane operator hoisted a massive anchor into the air and swung it out over the water up against the cable, so the cable passed right over the top of the anchor. A twenty-meter-long Kevlar pigtail was firmly attached to the anchor, and loosely coiled and tied to its top. Divers Wells, Solak, Kilker, and Apryl rode the anchor out over the side, with Bihm remaining on the barge to supervise. The crane operator lowered the anchor into the water, keeping it tight against the cable until the divers indicated they had passed the payout reel.

Under George's watchful eye, the divers quickly attached the pigtail to the cable just below the payout reel with a set of ten specially designed clamps. Then Kilker and Apryl swam twenty meters below the attachment point, where they attached a pressure-activated explosive cutter to the cable. Then the divers surfaced and were hauled in a steel basket to the deck where they were assaulted by a half-dozen lesser frigatebirds looking for food handouts. Once the divers were safely on deck, the barge operator swung the barge around the cable so that it was clear of the cable, and signaled the crane operator to drop the anchor.

Instantly, the dolphins disappeared to get a head start on the dropping anchor. At one hundred meters, the explosive cutter separated the lower part of the cable, which plunged on its own track to the bottom. Fifty seconds later, the 850-metric-ton anchor struck the bottom, driving its pylons deep into the sediment, followed by nine explosive releases of the anchoring cross-members. The dolphins surfaced with a flurry and crossed the distance to W-3 with remarkable speed.

The scenario repeated itself with minor variations at W-3, and when both anchors were firmly attached to the bottom, Margo got ready to make another bottom excursion in *Wampus*.

EQUATORIAL PACIFIC—AT W-1

RV Amelia E had brought Margo and *Wampus* to the drop spot over the W-1 anchor. By this time, it was late in the afternoon, but Margo wanted to complete the two anchor installation repairs as soon as possible so Ascension could continue. Pearl deployed *Wampus* over the fantail with Margo inside, and George and the other dolphins outside. The divers ran their external checks, and Apryl once again did her coquettish best to give Margo a fun sendoff. As she posed in front of *Wampus*, George came up behind her with a gentle push, causing her to tumble head over heels. Margo chuckled and dropped to fifty meters. The divers did a final check, and Apryl—after her manner—crossed her arms over breasts, signaling a safe journey. This time, George let her be.

Wasting no time, Margo dropped at full speed, leaving the dolphins at 300 meters, and reaching a hover near the bottom in eight minutes. She switched on her lights, quickly located the anchor, and drove herself to a position a couple of meters above the anchor surface. She located the bitter end of the cable and grasped it in her left manipulator. Then, with a smooth pushing motion, she threaded the cable through a formed channel in the anchor, and then lifted the bitter end to a point above the ten clamps installed at the surface by the divers. She attached the bitter end to the main cable with another ten clamps so that the cable was not only firmly attached to the pigtail that was attached to the anchor, but was also attached directly to the anchor. The entire task took just over a half-hour.

Margo examined her handiwork and reported to Klaus, "The repairs look pretty good, Klaus. I know you did the calcs, but only time will tell if it will hold like the original."

"Not to worry, Girl," Klaus said. "It'll hold."

Margo punched the coordinates of W-3. An hour later, she was hovering over the second anchor. She was disappointed to find the last two meters of the cable bitter end under the anchor. Cutting the cable free was the easiest solution, but would leave her with barely sufficient to clamp to itself above the pigtail. She grabbed the cable with both manipulators and commenced reverse thrust until she was at maximum. It didn't budge. Margo decided to loosen the cable with her waterjet, which consisted of a high-pressure pump with its output directed through a flexible hose to a narrow nozzle. She clamped the nozzle in her right manipulator and then activated the pump. As she directed the high-pressure stream to the bottom where the cable disappeared under the anchor, clouds of silt filled the water in every direction, completely blinding Margo. She turned off the pump and waited for the current to move the cloud away. When the water had cleared, she examined the point using a holocam and bright light. It looked like she had penetrated at least a half-meter. She set up again with the waterjet, this time pushing the nozzle as far as possible into the opening she had created.

When Margo activated the pump this time, she produced far less clouding silt. She kept the waterjet going for a full minute, feeding it into the hole. She stopped and let the cloud clear. Then she grasped the cable with both manipulators and built her reverse thrust up to full, and then let it off. The cable moved! She did it again…and again…and on the fourth time, the cable pulled free.

Klaus had been observing Margo's progress silently with his topside monitors. "Nice job!" he said as the cable pulled free. "Couldn't have done better myself."

The job was wrapped up twenty minutes later.

"Klaus, I'm off to see the wizard," Margo said. "I want to take another close look at *Electra*." She entered the *Electra's* coordinates, and fifteen minutes later, the outline of the aircraft took shape on her display, and shortly thereafter, she was looking once again at Amelia Earhart's actual final resting place.

With great care, Margo used her thrusters to "blow" the accumulated silt from the aircraft's nose. The nose and windscreen were smashed, and the cockpit was nearly filled with fine silt. Using her waterjet very gingerly, Margo commenced removing the soft, fluffy material. She cleared for a few seconds and then waited for the current to sweep the silt cloud away. Then cleared again…and waited…cleared again…and waited. As the cloud cleared, and as the silt-filled water slowly cleared the cockpit through the broken windscreen, Margo saw a shape taking place. As it came into gradual focus, she gasped and pulled back.

"Oh my God," she said to no one in particular as a skull began to take shape in the still cloudy water. It was wrapped tightly in a leather skullcap that seemed to be holding it in a very unnatural position, with strands of hair peeking around the edge over the forehead. Margo sat quietly, staring at the head, tears silently streaming down her cheeks. "Oh, Amelia," she said, "so this is how it happened." She looked more closely as the water cleared with a sudden small shift in current. "You hit the water with your right wing. Then you must have flipped, so your nose hit the water and broke the windscreen, and that, my darling sister, broke your neck. And then somehow, you ended down here, upright like this." Margo sobbed quietly, unaware of the respectful audience that had crowded into the Western Complex Control Center to watch the proceedings on the large holodisplay.

Carefully and respectfully, Margo moved around *Electra*, ensuring that her holocams recorded every inch of the aircraft and its surroundings. On a sudden impulse, Margo said, "Klaus, is Lori getting this?"

"Margo, this is Lori—every minute, every inch. My tears are mingling with yours, Girl."

Margo took her time coming to the surface. She wanted to be alone with her thoughts for a bit longer.

CHAPTER EIGHTEEN

BAKER ISLAND—OPERATIONS COMPOUND

With W-1 and W-3 back in business, Alex focused on completing Ascension. He turned over to Bruce the media storm resulting from Margo's discovery of Amelia Earhart's *Electra*. He needed to focus, and Margo's distraction was just that—an unnecessary distraction. Alex didn't begrudge Margo her moment of discovery. He well remembered the long evening hours over Merlot and single malt listening to her talk about her quest for Amelia, about Amelia's conquests, her highs and her lows. Alex had come to understand that Margo saw in herself many of the qualities she saw in Earhart. He had become very fond of this extraordinary woman whose own accomplishments in so many ways eclipsed those of her heroine.

Alex turned to Lori, who hovered nearby. "Might as well pack it in for the night, Lori," he said. "It looks like the excitement is over. We've got ten days of boring routine ahead of us now." Then he turned to Bruce, who appeared to be none the worse for his extraordinary ordeal. "Bruce, please clear the media people from the room and get them settled in for the night." Finally, he placed a Link call. "Mabel, it's Alex. We have everything back under control. Ascension is underway again."

"Do you think this is the last? Do you expect something else?"

"Sorry, Mabel, answering that question is beyond my paygrade. We lost the little sonofabitch in the struggle. He's making tracks for Pago Pago. I suppose you could pick him up there, but we don't have anything that would hold up in court, and I don't have anyone to spare." He paused. "You could always try to work something against his sponsors, you know."

"We're working on that one, Alex." She smiled at him. "You look tired. Get some rest." She terminated the Link.

SEATTLE—SMITH TOWER

"So what happened?" Quinton Radler asked angrily of his assembled team three days later in the conference room on the fourteenth floor of Seattle's Smith Tower. "Why is Ascension still happening?"

"It was a good plan, Quinton," Gotch said. "Watson's timing may have been a bit off—it seems he got to the underwater site later than planned, so he attached only two explosive cutters instead of four. And," Gotch added, "he got tangled on the bottom. One of Slingshot's super minisubs rescued him."

Radler looked out the window opposite him. It was one of those rare beautiful Seattle days when you could almost see forever. The Space Needle was sharply outlined against a cloudless blue sky. "Can we rescue this?" he asked.

"They've got five pairs of tensioners on each downslope and three pairs along the rail," Gotch said, "and they have boats parked at each." He spread his hands. "Unless we do a total underwater approach, we're not going to get another bite at this part of the apple."

"Doing that's outside our cost-benefit envelope anyway," Radler growled. "Let's move to the next phase."

AMERICAN SAMOA—PAGO PAGO

Watson arrived in Pago Pago nearly recovered from his ordeal. He was upset and angered by what had happened, and still did not understand how he had gotten the timing so wrong. He figured he was lucky still to be alive. After the accident, his special Link connection had ceased working, but when he stepped off the *Aku Aku*, a stranger slipped a note into his hand. The note instructed him to go to a public Link and call an encrypted connection. When he did that, he was instructed to go to another public Link from which he was sent to the outside seating section of a small restaurant. While sitting at the table, sipping a drink, a man approached him with a set of written instructions. Watson couldn't tell if it was a different person than the man who originally gave him the note as he stepped off the ship, or the same person dressed differently. In any case, the note gave him another encrypted connection that would work from his personal Link. Watson had become accustomed to the strange methods he had to follow when dealing with his benefactors; he had concluded that they must be prominent, and wished to avoid any recognizable connection with *Green Force*. He was determined, however, to pump them for all he could before they went away.

The new instructions Watson received from his reestablished encrypted connection were interesting. He was to take a "leave of absence" from the directorship of *Green Force*, and put one of his trusted deputies in charge. He was to tell that deputy nothing. Then they presented him with an image of himself with short dark hair, and dressed in conventional business attire. He was instructed to make himself look like the image. A barbershop was just across the street, and a men's clothing store was a block up the street to the right.

BAKER ISLAND—OPERATIONS COMPOUND

Alex did not sleep much, despite Mabel's admonition. Simply stated, he was keeping track of too many things. He had complete faith in his senior people, and he knew their people were handpicked for their skills and loyalty. In the final analysis, however, he was *the man*. This phase of Ascension was delicate and fragile. It had never been done before. Each problem was a new challenge. The designers had exhaustively tested their models against every known factor. Virtually every conceivable event had been examined from all possible aspects. Despite all this, nobody had anticipated the *Green Force* fleet. Nobody had foreseen a minisub cutting tensioner cables near the seafloor anchors. How many other things, Alex wondered, lay before him that nobody had anticipated, foreseen, or planned for? He was only too aware that Mabel looked to him for the leadership that kept the project going, and everyone else on the project looked to him for guidance and oversight, relying on him to correct their mistakes. Alex was a man filled with self-confidence—otherwise, he would not have been in his position—but he was very much aware of his own shortcomings. He carefully masked his own feelings, and presented a strong, confident face to Mabel and to those beneath him. He was too honest to kid himself, however.

So, Alex did not sleep much.

Alex was particularly unsatisfied with the deployment of the tensioner cables. The massive payout reels somewhere in the water column simply didn't make sense in actual practice, although they must have performed well in the simulations. He was well aware of the opposing design camps. One wanted to use large barges outfitted with extra-heavy-duty station-keeping thrusters at each station. During Ascension, each barge would feed cable and maintain stability of the loop with its thrusters. When Slingshot was aloft, each barge would drop its tensioner anchor and let the automatics take over. The other camp, the one that prevailed, wanted to put the anchors and payout reels in place before Ascension. Although there were all kinds of technical reasons for one approach over the other, the prevailing camp won out primarily with their argument that the anchors married to the bottom before Ascension provided more stability than barges floating on the surface, even barges with super-massive station-keeping ability, AND the cost was a small fraction of the cost for twenty-six of the proposed barges. Alex didn't buy the first reason, but he understood only too well how cost influenced everything he did.

Alex consoled himself with the thought that Slingshot was a proof-of-concept project. They had chosen the ideal location, the best design—so far as they knew, having never done it before, and the most cost-effective construction techniques they could devise—with the same proviso that it had never been done before.

The process seemed to be working well. The rail bootstrapped itself skyward with a deliberate pace as the two ocean-going tugs pushed eastward and westward respectively on the great plates at the apexes of the Eastern and Western Complexes. As the two downslopes lifted out of the water exposing the tensioner attachment points, ten payout reels fed cable in a delicate dance of balance—counterbalance, keeping the downslopes and rail perfectly plumb. As the rail rose, three pairs of tensioners extended themselves from the abyssal plain five kilometers below the surface, evenly spaced along the length of the rail, maintaining the rail's verticality. The process was entirely monitored by computer, since reaction time to any unforeseen event was time-critical. By the time Alex or one of his people would have recognized a problem and taken corrective action, Slingshot would have torn itself apart.

The one aspect of Ascension that required the closest human monitoring was the ongoing construction of the skytowers. Stripped to its essence, a skytower consisted of a skytower cable, a suspensor cable permanently anchored to the skytower cable at points all along the cable, and a double lift cable that passed through spacers attached to the suspensor. The actual lift cable consisted of two parts, a high-speed three-gee boost cable and a one-gee lift cable. The boost cable accelerated the capsule at three-gees for eight kilometers, handed it off to the lift cable for a one-gee acceleration phase for sixty-four kilometers, with a final upward coasting freefall for the last eight kilometers. While the system was elaborate, with its double driving mechanisms, one to supersonic velocities, overall, it still consisted of a supporting cable and an elevator with a pulley-supported cable system. This system distinguished itself by being able to lift a capsule from the ground to the skyport every five minutes or so. This made the launch loop commercially viable.

In actual practice, the skytower cable was much more than a simple Kevlar cable hanging from an attachment point at the skyport, called the terminus, and extending to the socket on Baker. Since the skyport was subject to measurable horizontal motion and even some occasional vertical movement, the cable end had to be firmly anchored, but in a way that allowed for virtually frictionless angular movement and occasional vertical excursions. Furthermore, rising capsules had to be constrained so they did not twist around the cable. This was accomplished with gyroscopic stabilization for each capsule.

The skytower drive consisted of a linear driver on the ground to drive the boost cable, with a coupling device at eight kilometers that transferred energy from the three-gee boost cable to the one-gee lift cable. The cable design for both cables was an exotic combination of Kevlar fiber infused in the manufacturing process with a matrix of neodymium and several of the lanthanide series of elements, carbon nanotubes, and steel wire. The resulting cable was magnetic and very much lighter than steel, with many times its tensile strength. Because of its permanent magnetic character, it could be accelerated with a linear driver, and suspended in a frictionless motion deflector instead of using traditional bearing-based pulley wheels.

As Slingshot continued to bootstrap itself skyward, teams on Baker and Jarvis performed virtually identical actions. Initially, they loosely passed the skytower cable through the anchoring loops in the sockets. As the rail moved up, they fed cable from a huge reel located on a barge in the harbor, keeping it loose, without tension. Simultaneously, they attached the lift-cable suspensor to the skytower cable every five hundred meters with aramid-based polymer rings. Both the lift and boost cables passed through meter-long tubes connected to each ring that were lined with neodymium magnets. These tubes restrained the cables without friction. The eighty-kilometer suspensor cable with its attached rings and tubes and the one-gee and three-gee cables were married together during the manufacturing process. While this guaranteed the integrity of the assembly, it complicated the installation process in a major way. Whereas the main cable was simply wound on a large reel and fed through the anchoring loops in the sockets, the combination of suspensor and lift cables was clumsy and awkward to feed to the rising rail. It could not be wound on a reel nor be attached to the main cable and fed through the anchoring loops, because this would damage the magnetic tubes. Consequently, the married suspensor and lift cables were loosely laid in two large barges tied up to Meyerton Landing Wharf at Baker, and Millersville Wharf at Jarvis. A crane on each wharf suspended a large-diameter cushioned reel over each barge, and A-frame structures held twelve similar reels starting at the berms above the two wharfs, one every hundred meters for the one-and-a-quarter-kilometer distance from the wharfs to the sockets. Cushioned windlasses at the sockets gently pulled the married suspensor and lift cables from the barges, across the reels, to the center of the sockets where they were attached to the skytower cables. The skytower cables themselves crossed a similar set of A-frames, but these were outfitted with simple rollers that facilitated pulling the cable off the reels.

In the Baker Operations Center, Alex monitored this critical phase. Every five hundred meters of elevation, the computer controller stopped the Ascension for the installation of the rings and their associated tubes and cables. During the half-hour or so that this operation required at each socket, the ocean-going tugs had their fuel topped off, and people throughout the entire project inspected specific components and reported their results to Alex. Alex knew that overall control was continuously monitored by his computer system and its redundant backups, but in the final analysis, Alex was convinced that success or failure was people-dependent, not computer-dependent. When the shit hit the fan during the Ascension, Margo pulled their fat out of the fire, not the computer system. Alex understood full well that the computer controller caught the initial problem and stabilized the system, but Margo did the repairs.

The rise, stop, rise, stop process continued for nine long days. Eight kilometers from altitude, Alex paused the rise to install the frictionless motion deflector at the top of the boost cable. This is where the boost cable turned back on itself and headed down to the ground, and where it transferred some of its energy to the lift cable. The installation was tricky, because this was also where an upward-moving capsule transferred from the boost to the lift cable for its sixty-four kilometer trip to the next decoupling point. From there, the capsule would coast the final eight-kilometer distance to the skyport, arriving at the top of its apogee with near-zero velocity.

The installation of both frictionless motion deflectors took an entire day. Upon testing the Baker installation that evening, the computer controller reported a significant likelihood that upon activation, the frictionless motion deflector would rotate about its axis, and consequently around the main cable. Alex and Klaus linked together and conducted a manual analysis of the potential problem.

An hour or so later, Klaus said, "It looks like the computer is correct. The original model analysis missed the effect of these stresses," he pointed to several terms in the analysis, "and so did not predict any significant torque."

"I agree, Klaus, so what do we do? We can't send it up, and we can't bring it down. We need an in-place solution."

"Let me think about that, Alex. I know you need to report to Mabel, but give me an hour," Klaus said. "I have an idea."

WESTERN COMPLEX—CONTROL CENTER

In the Western Complex Control Center, Klaus called up one schematic after the other, looking specifically at the bottom and sides of the frictionless motion deflectors and at how the capsule gyrostabilizers were attached. Although Klaus enjoyed posturing with his rolled paper schematics, there were times when he thoroughly appreciated the modern digital world. This was such a time. Since the system had access to every component, piece of equipment, and spare part, Klaus figured that it might be able to jury-rig something together that would be a permanent fix, or at least last sufficiently long to enable Mabel to design and manufacture a real replacement.

Within several minutes, the system holodisplay presented a solution that incorporated a spare set of neodymium tubes to extract power from both legs of the moving cable just below the frictionless motion deflector, and a simple, secure four-point attaching interface to join it to the stabilizer. The interface was a modified cylinder-head block from his pickup spare parts. It was brilliant! Klaus called Alex.

"Look at this, Alex. We're talking an hour—maybe ninety minutes to make each one. The computer says it'll counter any torque the frictionless motion deflector can produce."

"You're too damn impressive, Klaus. You got to stop doing things like this," Alex said with a face-splitting grin. "You get them produced while I brief Mabel."

SEATTLE—SMITH TOWER

In Seattle, high atop the Smith Tower, Mabel sat at her ornate desk contemplating the holodisplay in the air over the desk. The sun had set in a cloudy sky, but the city lights filtering up from Pioneer Square and across from the taller buildings scattered around her gave her a sense of belonging to a living cityscape, enhanced by the glowing flames from the gas fireplace on one side of the room. The holodisplay over her desk duplicated on a smaller scale what Alex had on display in the Baker Operations Center. *Alex, my lad*, she thought, *you are my best hire ever. I know you have Klaus and Margo, but you bring it all together. You are the apple of my eye down there.* She felt something akin to a mother's pride. With no children of her own and zero prospects of that ever happening, Mabel was enormously proud of her people. Immediately after Alex's briefing, she had put into motion the design and manufacture of replacement frictionless motion deflectors. At best, it would take two or three weeks. It could take, she knew only too well, a month or more—even longer if the mysterious *EI* got involved with the design and manufacturing process. *Alex, you pulled it off in three hours!* She wanted to open her permanently sealed windows and shout it to the housetops. *Damn, I've got a good crew!*

At this moment, however, all she wanted to do was lurk in the background to see how her "kids" were doing. She had always thought that ultimately, Alex and Margo would get together, but in the past few weeks, she noticed that the bond between Klaus and Margo was getting stronger. She had watched with amusement the interplay between that reporter, Lori, and Margo. Mabel suspected something was afoot but didn't probe any deeper. She had become enormously fond of Bruce and was particularly delighted to see him step up to the plate. The other youngsters from *Green Force*—her admiration for how Alex had handled them knew no bounds. She had told the story to every executive she knew, challenging any of them to come up with anyone so capable as her Alex!

BAKER ISLAND—OPERATIONS COMPOUND

Two days later Alex transmitted on All-Call and to every loudspeaker in the entire project: "Ascension is complete! Slingshot is stable. Space—here we come…*Ad Astra!*"

CHAPTER NINETEEN

SEATTLE—SMITH TOWER

When the weather was clear, Dane Curvin had a ringside view of Mt. Rainier through his eastern window to complement his view of the lush interior of the Smith Tower well through the south window opposite his desk. Today was such a day, and the snow-covered slopes of the volcano dominated the eastern skyline. Curvin never tired of the view. He Linked to one of his fellow Yale graduates, in-house counsel for one of the many thousands of suppliers to Slingshot.

"John, I need a favor," Curvin told him.

Curvin then set up an arrangement for Lars Watson to be hired by Slingshot as a high-iron worker under the assumed name of Bob Weingard, Curvin's "third cousin." *EI* would ensure that his background record reflected the necessary skills and experience.

"No, John," Curvin explained, "it's not like construction in orbit. You're weightless in orbit. This is like building the highest skyscraper in the world. You fall from a skyport, you got an eighty-kilometer drop! But you still wear spacesuits…no atmosphere up there." Curvin clarified the details and disconnected the Link. He leaned back in his chair, hands clasped behind his head, admiring the eastern view. *It'll be interesting to see exactly how Gotch sets up the skyport sabotage,* he thought. *I wonder if that idiot Watson realizes what will happen to him if he's still up there when Slingshot crashes?*

HOWLAND AND BAKER ISLANDS

Following six weeks of specialized training on the specific techniques for constructing the skyports, Bob Weingard, a.k.a. Lars Watson, flew into *Amelia Mary Earhart International Airport* with his hire group—twenty-one high-iron guys from all over the world. Five were Mohawks from Brooklyn who had trained with him at the Seattle facility. Watson had never given much thought to how tall buildings and bridges got built. From his perspective, they were just there, objects that were part of existence. But he did know, as did most everybody else, that Mohawks had been high-iron workers for nearly ten generations. He wasn't entirely certain why traditional high-iron workers like the Mohawks would be hired for construction in space, but he figured the guys in charge probably knew what they were doing. Watson intended to learn as much as possible during the next few weeks so he could cause his mischief in the most effective way when the time was ripe. This time, he promised himself, he would not be "late" in coming to the table. He would be present and accounted for when the time for action arrived.

Watson had been hired as an inspector. His job was to inspect rivets, welds, and the new molecular bonding techniques that had come out of the past fifty or so years of space construction. Gotch could not have placed Watson in a position where he had to perform a specific skill like welding, riveting, or bonding, since Watson had no such skills. But an inspector's knowledge is presumed, so Gotch arranged for Watson to be hired as a general inspector. He sent Watson to a training course so he would be familiar with the terminology and the equipment he would encounter at the skyport. Since Watson's ultimate task was to destroy Slingshot, his level of knowledge was irrelevant, and the accuracy of his inspections completely irrelevant.

Watson found himself paired with Peter LaFleur, who turned out to be a thirty-two-year-old distant grandson of his namesake—the

1960s photographer and high-iron worker who has given the modern world an intimate picture of many of the great high-rise projects of the mid-twentieth century. LaFleur had worked literally the world over on many of the world's highest projects. Watson liked him, and had to restrain himself from trying to convert the Mohawk to his savage brand of environmentalism. *Your only job*, his encrypted link had insisted, *is to bring down Slingshot*. Watson resisted the temptation and restricted himself to small talk as they were processed through security.

A half-hour later, Watson and LaFleur boarded a new-model Chinook with the remaining nineteen team members for the short trip to Baker. Upon landing on the tarmac on the south side of the island, they were loaded into a small bus and driven to the Processing Center at the base of *Baker Skytower*. The center was still under construction, which meant there was hardly any formality with their secondary security inspection. It amounted to little more than a cursory glance at their identification before they walked through full-body scanners on their way to outfitting and final briefing.

Watson looked around curiously at the half-finished building, a coral-white one-story structure with windows on all sides, an entry on their side, and an exit leading to the capsule boarding ramp at the base of the skytower. Although it was still early morning, the sun was already in full bloom in a sky littered with a few fleecy clouds and perhaps a couple thousand birds that he didn't recognize. Although Watson was an avowed environmentalist, he saw himself as an activist protecting the Earth, not as someone who needed to know details like the names of specific birds in the sky. He knew that sooty terns, masked boobies, and lesser frigatebirds were native to Baker, but he could not have pointed one out if his life depended on it. He knew that Slingshot had endangered all of them out there, giving no thought to their increased numbers since the commercialization of Baker. *Lies*, he told himself, *lies to fool the public. But not me*, he insisted, letting a little smile creep across his face. *Not me!* He and LaFleur strolled into the cool building interior.

"Ah…that's better," Watson said to LaFleur as they settled into utilitarian chairs in the briefing room.

"*Khena'tonhkwa Pussy!*" LaFleur responded with a wide grin.

"Kena…what?"

"I said, Weingard, that you're a Pussy in pidgin Mohawk." Adding, "Can't take the heat, get outa da kitchen."

BAKER ISLAND—OPERATIONS COMPOUND

Bruce Yoon strode nervously to the podium facing the small crowd seated in the temporary assembly room in the Processing Center. He was keenly aware of his youth as he looked out at the twenty high-iron veterans—all of them tough and wiry with more years on high-iron than he had been alive. Behind them and to the sides was a small crowd of media types augmented by any local Baker personnel who had nothing better to do at that moment. As he stood, collecting his thoughts, Lori caught his eye from the back and winked. She was stunning in loose white slacks and a modest white blouse that somehow screamed her femininity despite its conservative cut. Bruce blushed slightly, glanced out the window to his right as the morning cacophony of sea bird noises filtered through the insulated window. He glanced down at his Link display that he had set to show his prepared words on the podium surface, and cleared his voice. He looked up to see Alex, Margo, and Klaus stroll into the room. Alex gave him a thumbs-up. Bruce cleared his voice again and activated a holodisplay that covered the wall behind him.

"We don't stand on ceremony here," Bruce commenced. "You guys will be the first humans in history to travel into space by elevator." He paused to let that sink in as the audience murmured. "The first," he repeated. "NASA awards astronaut status to those who fly fifty miles high, you know," he added.

Then Bruce went into detail, describing the fundamental nature of Slingshot, with appropriate holodisplays behind him. He discovered that as he got into his presentation, he really didn't need his prepared script. He knew the project inside out, and except for using his link to remember the order of his presentation, he found himself doing it extemporaneously. From time to time, he glanced at Alex, who beamed like a proud father, and at Margo and Klaus, whom he thought of as older siblings. He purposefully avoided Lori's eyes, knowing that she would cause him to blush again.

Bruce described the compact package at the intersection of the rail and the downslope that had hitched a ride all the way up during Ascension. It contained, he explained, an extensible tube that, after arriving at the eighty-kilometer altitude, had pushed itself upward, propelled by compressed gas. Once it reached its twenty-meter height, it had hardened to a consistency stronger than steel from its exposure to the sun's UV rays. As it extended upward, a ball at its tip had deployed ten hybrid carbon nanotube-aramid fiber cables attached to the ends of ten horizontally extensible tubes at the base of the vertical tube just two meters above the rail. Compressed gas extended the tubes fifteen meters outward in a horizontal array fanned five to the north and five to the south, supported by the gossamer cables. As with the vertical tube, the ten horizontal tubes hardened under the influence of the sun's UV.

Bruce explained that the crew had been divided into three shifts. He read off the names of the first two. Both Watson and LaFleur were in Shift One. Their designated Shift Boss was another Mohawk, Mike Swamp. Swamp was a thirty-five-year-old, 178-centimeter tall, direct-line descendant of the mid-twentieth century Mike Swamp of New York skyline fame. Besides English, he was fluent in Mohawk, and comfortable in a half-dozen other languages as well.

"Those of you I haven't called," Bruce said, "are in the third shift, of course."

He explained that Shift One would ride to the node first while Shifts Two and Three would load up the freight capsule with the initial construction materials. He told them they would send up the freight capsule as soon as possible after the passenger capsule had returned to the socket, and that they would send up three additional freight loads. Eight hours later, he said, Shift Two would be sent up, and Shift One would return in their capsule.

"Project boss Alex Regent will now say a few words," Bruce said at the end of his presentation. "Then you'll get dressed out and load up in the capsule." He smiled broadly at the assembled high-iron crew. "Good luck, guys! I wish I were going with you."

❋

Alex sauntered to the front of the room, looking over his Shift One crew, catching an eye here and there. He could hear the avian cacophony through the windows as he passed, and when he glanced outside, it almost seemed as if someone had spread food on the ungroomed surface surrounding the compound. The birds seemed to know something was up.

"A couple of hours from now," Alex said as he reached the front of the room, "you guys will be higher than those birds will ever go." He pointed out the window to his right. A nervous laugh rippled through the group.

"I want to walk you through your shift," Alex continued. "The third leg of your ride will be at freefall." He pointed to the holodisplay behind him, where an animation showed the final stage of the ascent. "You're going to feel weightless during the first part of this leg, and then weight will gradually return. As you reach the top of the skytower, your full weight will return as your capsule clamps to the skytower cable at the terminus, right here." Alex indicated the point where the rail, downslope, and skytower intersected. "Your capsule will stop directly below the fan Bruce just briefed you about. Now here's the tough part—and probably the most dangerous." Alex paused to allow the holodisplay to give an animated close-up. "You are going to attach a rope ladder to the nearest fan brace, and climb the ladder one at a time to the top of the fan, where you will attach yourselves and await the arrival of the first cargo load." Alex disabled the holodisplay. "Remember, the ground is eighty klicks below you. Until you install them, there are no safety nets."

※

Watson felt confused. From what he understood about orbital operations, you were weightless all the time. He resisted the urge to ask the obvious question. The last thing he wanted was to call attention to himself. *I got to think about this, but there's no time. We're about to get suited up. Then the ride to the skyport, and then…* "Let's go, Peter." He stayed close to his Mohawk friend as they filed out of the meeting room and down the hall to suit up.

Unlike older vacuum suits, the suits issued to Watson and his six companions were lightweight and flexible, even in hard vacuum. They worked on a technologically advanced application of earlier high-altitude suits. An inner garment formed a flexible, skintight membrane that substituted for atmospheric pressure. A slightly less tightly fitting outer garment retained a minimal atmospheric environment inside the suit, and provided temperature and wear resistance. Watson and the others removed their footwear and outer clothing. They entered the suits feet first through an airtight zipper-like opening in the suit's back. As he removed his shoes, LaFleur remarked, "What's the red and green sock routine? I bet you got another pair just like it in your sock drawer." He laughed. "You color blind or sumptin'?" Watson growled at LaFleur, saying nothing.

For Watson, the suit took a bit of getting used to. At first, it felt like his chest was being squeezed. When he breathed deeply as he was instructed to do, the pressure seemed to relax, although it came back as he exhaled. He found that with a bit of effort, it almost seemed like the suit was helping him breathe. The gloves were comfortably flexible, although Watson decided he would be glad to remove them when he returned to normal atmospheric pressure. He pitied the guys who had to flex their fingers for eight hours. They each were outfitted with a close-fitting skullcap that contained, they were told, various sensors that would transmit their physiological condition through a booster on the skyport to the Earth-side monitors at the socket. Watson slipped his head into the transparent, spherical helmet his outfitter handed him, and pressed it to the sealing collar around his neck, twisting it several centimeters to the right just like they had showed him. The collar sealed to the clear globe with a faint click; from inside, the helmet was completely invisible. Watson felt a faint movement of cool air as his breathing system cut in. Simultaneously, Watson's suit Link activated. A holodisplay appeared in front of him exactly as it would under normal conditions using a regular Link. Watson had been fearful that he would feel claustrophobic inside the globe, but with it on and sealed, he shortly forgot it was there.

"Comm check, Weingard." It took Watson a moment to react. "Weingard, Comm check!" Watson reached up to adjust his skullcap, forgetting about the helmet, and struck it with his gloved hand.

"This is Weingard…roger." He still was getting used to his new name.

Watson examined LaFleur's backpack as their outfitters had instructed. It consisted of three relatively thin cylinders. Two were high-pressure gas tanks that held pure oxygen. The middle slightly larger one contained a breathing bag that scrubbed carbon dioxide with the same electronic-molecular trick used by *Wampus* and *Wallus*. The gas fed through a flexible tube attached to the suit's right shoulder into the suit just below the collar ring under the chin, and back out through a similar tube over the left shoulder. The electronics pack nestled below the cylinders at the small of the back. After Watson finished his examination of LaFleur's backpack, he attached the lightweight flush fairing as the outfitter had demonstrated. The fairing served to eliminate possible snag points on the backpack. The fairing displayed in bright fluorescent letters: LAFLEUR, although to Watson, they appeared only as a bright white.

Following Watson's indication that his unit was okay, LaFleur inspected Watson's backpack and installed his fairing with its fluorescent label: WEINGARD.

Watson placed his gloved hands on both sides of his helmet and rotated it to the left several centimeters. With a soft hiss, the seal released, and he removed the clear globe. LaFleur did likewise, and they tucked their helmets under their arms and followed the other crewmembers out the door into the oven-like heat of the late morning Baker sun.

Although the suits were fully reflective, Watson still could feel the sun's tropical warmth beginning to penetrate his suit. The outfitters had told them that in another week or so, they would be able to go from outfitting directly to the capsule ramp without going outside, but that wasn't quite ready yet. Watson strode alongside LaFleur with a degree of uneasiness that he was certain was visible to everyone who looked at him. He walked without comment, remembering to keep a low profile.

Ahead was the capsule. It looked like a short, shiny metallic sausage. It had four gull-wing doors on the port side, each with a large window, and the nose cap opened on a hinge at the top. The starboard side had matching windows, but no doors. As the crew approached the capsule, all the doors opened. Slingshot staff member Rudolph Pigman waited for them, his 173-centimeter stocky frame suited up with clear helmet under his left arm. Although his shaved pate was covered by his skullcap, his forehead glistened in the late morning sun. "I'm your flight attendant for this trip," he announced. "Sorry, but we got no refreshments, no peanuts, no nothing on this flight, and I ain't wearin' a skirt!" A wave of nervous laughter swept through the waiting shift members. "Mike Swamp," he said, "you get to go down in history—you'll be the first man to exit the capsule onto the skyport. An honor for the Mohawk nation. Sit with me in the front row. Weingard and LaFleur, you're next. The rest of you follow. Strap in tightly. I mean that—tightly. Each seat is equipped with an ejection device and a parachute. If something goes wrong, you won't even have time to think about it. One moment you're here," he indicated the capsule, "and the next you're floating down through the sky, hoping like hell someone's gonna find you when you hit the drink." Nervous laughter again. "Don't worry about it! Each seat and your suits have tracking devices. We'll find you alright." He grinned at them. "We'll find you alright. Now, get in, buckle down, and don your helmets."

Watson walked to the second gull wing and stepped into the capsule. His seat was specifically designed to accommodate his backpack and helmet. He nestled himself into the seat and felt it cocoon around his back and equipment. He secured his safety straps across both shoulders and lap, remembering Pigman's admonition, donned his helmet, and sat back to await the launch. Once the rest of the team were installed, the gull-wing doors and the nosecone closed, and shortly thereafter, the capsule moved forward several meters. Through their bone conduction phones, Watson and the others were told to relax while the capsule was hoisted to its vertical lift position. Watson heard a distinct clank, and then the capsule slowly pitched upward until he was on his back with his legs and feet above him.

"Keep your legs together in the cushioned indentations," Pigman told them on their common Link. "Keep your arms in the armrest indentations." He paused. "Relax, guys. This is no big deal. The rocket boys pull more than twice the gees you'll pull on this lift—they get off on it!"

Then a virtually featureless voice that had to originate in a computer droned a short countdown: *ten…nine…eight… …three… two…one…*

Watson felt a faint tug that rapidly increased to about the same pressure as one of his environmental chicks sitting on his chest. He could just hear the high-pitched whine of the gyrostabilizer, and he heard an extended whoosh and turned his head to see fleecy clouds whip past his window. Seventeen seconds later, and eight kilometers above the ground, Pigman announced over their common Link: "We just passed the sound barrier, Boys!" A few seconds later, the weight on Watson's chest decreased slowly until he felt completely normal. Back to one gee, he told himself. The capsule shuddered, accompanied by a clank as the capsule shifted from the boost cable to the lift cable. The whooshing faded as the capsule, rising at more than 400 meters per second, left most of the Earth's air behind as it entered the rarified air of the upper atmosphere. There was no vibration, no sense of movement at all—it was as if they were still on the ground. Watson looked out his port, but all he saw was an increasingly dark sky.

The sensation of being on the ground seemed to last forever, but it actually lasted for only three minutes. At seventy-two kilometers above the ground, the capsule released its hold on the lift cable and commenced its final eight-kilometer freefall to the incipient skyport. Watson's initial reaction was panic as all weight disappeared. His body knew he was falling. Despite his intellect, every fiber in his being shouted *Falling…Falling!* His gorge rose in reaction. He swallowed hard to quell his stomach, and gripped his armrests fiercely. He stole a glance at LaFleur across the narrow aisle. LaFleur's arms floated in mid-air, and the Mohawk had a blissful look on his face. *Damn you, Peter…* Watson fought to keep from hurling his stomach contents into his helmet. After about a minute-and-a-half, Watson heard a clank as the capsule clamped to the frictionless motion deflector connecting the lift cable to the incipient skyport. He felt weight return, and his nausea immediately disappeared.

"Check your helmets," Pigman said. When his Link indicated seven tight seals, he activated a control releasing the capsule's air to the outside. A moment later, the nosecone swung away, and Pigman stood up. "Welcome to *Amelia Earhart Skyport*," he said with a flourish. "Mr. Swamp, if you please…"

CHAPTER TWENTY

BAKER ISLAND—OPERATIONS COMPOUND

"You did great, Bruce," Alex said as Shift One filed out of the meeting room.

Lori ran up and gave Bruce a moist kiss. "You were wonderful, Sweetheart! Just wonderful!"

"Actually," Margo chimed in, "I agree."

"So do I," Klaus added.

"Gee, guys…" Bruce was pleased beyond words.

Alex put an arm around the young man's shoulders. "We'll get you up there as soon as the guys install some railings and safety nets. That's a promise." He guided Bruce through the door. "Let's go monitor the first lift."

Dex Lao slipped ahead of them and walked backward, recording their progress to the dressing room. Lori supplied a running commentary. For the next hour, Alex and the others mingled with the Shift One crew as they dressed out and tested their suits and helmets.

"Is it Mike Swamp?" Lori said to the big Mohawk. He grinned at her, reacting as did every man to her projected femininity. "I'm Lori from the Fox Syndicate." She pivoted herself and Swamp, so they partially faced Dex's holocams. "I understand you're the gang boss for Shift One? May I call you Mike?" He answered with a silly grin. "Well, Mike, how does it feel, as a member of the Mohawk Nation, to lead this crew on its historic ascent?"

"It's historic, alright," he said. "We—there're five of us, you know—we just want to live up to our ancestors' proud accomplishments."

"I understand you have a couple of team members with distinguished ancestors."

"That's right, Lori." The Mohawk's eyes twinkled as he spoke her first name. "Pete LaFleur was a high-iron man and well-known high-iron photographer in the mid-twentieth century, and my namesake ran the union and worked on every tall building in Manhattan right through the end of the century. All of us go way back."

"That's an amazing tradition—and now you're here, to work on the highest structure ever built." Lori smiled up at him, pouring on the charm.

"That's right. We couldn't be more excited." He glanced around the room at his crew, ready to leave. "I gotta go." The Mohawk took her right hand. "*Enwa:ton ken?*" he asked, and gallantly brought it to his lips. Eyes twinkling, he released her hand, adding, "*Nia:wen*."

"And what was that?" Lori asked, slightly flustered as she signaled Dex to stop the recording.

Swamp laughed. "A French custom married to Mohawk words. I said, May I? And then I thanked you." He grinned. "No embarrassing words for someone to figure out, I promise."

Lori lifted herself to her maximum height and kissed the big Mohawk on his cheek. "Sweet man," she whispered.

❋

Bruce watched in awe as the Shift One guys took their places inside the capsule. A high-tech version of the mechanism that pulls cars through a car wash moved the capsule along its track, and then pitched it to vertical. He shifted himself around so he could see the great clamps on the belly of the capsule. From his vantage point, they seemed to be gripping the boost cable, but he knew that couldn't be, because the boost cable was moving at supersonic speed.

"Stand back, everyone!" the launch boss ordered. "If something goes wrong, you don't want to be in line with the boost cable. Please stay within the cordoned-off area." He checked his link. "We're go!" he announced.

Bruce was nearly as excited as he would have been had he been inside the capsule, as the countdown commenced: *Ten…nine…eight… …three…two…one…*

To Bruce's utter astonishment, the capsule simply disappeared. One moment it was there, the next, it was gone. One moment he was concentrating on the clamps, the next, all he could see was the cable, or more properly, the blur that was what he could see of a cable moving at 400 meters per second.

The countdown switched to a count-up: *One…two…three… …fifteen…sixteen…seventeen…transfer to lift cable.* The count-up stopped, but every fifteen seconds, the disconnected voice announced the next quarter-minute, until it reached three minutes…commence freefall. The quarter-minute count-up continued until the voice announced, skyport clampdown.

Bruce glanced at his watch. A total of five minutes had passed.

※

"Okay, guys, gather 'round," the lift boss said to the fourteen assembled Shift Two and Three members. The passenger capsule had just returned and been carted to a sidetrack. In its place, a freight capsule rolled up and opened its clamshell top. "We're going to pack that stuff," he pointed to a stacked pile of carefully rolled flattened tubes, "and that stuff," he pointed at a pile of compact cylinders, "and that stuff," another pile of assorted tools and tubes of adhesive, "into that." He pointed at the freight capsule.

The crewmembers gathered around the freight capsule. It was divided into a series of slots that started at the back and extended all the way to the front. "Remember," the lift boss told them, "the freight will be subjected to three gees, and ultimately to zero gees. You need to pack it so that it will not get squished, and at the same time so that it will not come loose and damage something. You guys think you can handle something like that?"

Lori sidled up to Alex and said, "Alex, dear, may I borrow Bruce for a while? I want him to tell me all about the things those guys are loading."

Alex grinned and handed Bruce off to her. "Go for it, Lori." He winked at Bruce. Lori slipped her arm through Bruce's, and snuggled up to him as she steered him to the first pile of rolled flat tubes.

"So tell me, Bruce," she said, gazing at him through lowered eyelashes, "exactly what are these?"

Bruce blushed a bit but didn't extract himself from her closeness. "Well," he said, "you know how I described the vertical support tube and the ten horizontal tubes?" She nodded with slightly parted lips. "These are sorta the same thing. You can think of them as rolled-up floor planks. When they arrive at *Earhart Skyport*, Shift One will inflate them from those cylinders," he pointed to the stack of compact cylinders. "Once they unroll to flat, they will quickly cure into harder-than-steel light-weight planks. The guys will glue them to the ten fan elements using the high-tech adhesive in those tubes over there," he pointed to the third stack. "They interlock, so they can be securely attached to each other as the crew progresses across the fan." Bruce activated his link and projected a holodisplay. "Look at this," he said.

He showed Lori a cut-away image of the deck, and then showed her how several "planks" would be laid on edge across the bottom layer. These units, he pointed out, contained lightening holes that were designed to interconnect the individual spaces that would be formed when a third layer was assembled on top of the cross members. The sides and top would be similarly constructed, Bruce explained, and eventually, the space between the layers would be filled with a material that would expand into radiation-absorbing insulating foam.

"You really know Slingshot," Lori said, squeezing Bruce's arm. "I can see why Alex gives you so much media responsibility." Bruce's face turned crimson, but his eyes glowed. She leaned over and whispered in his ear, "Suck it in, Kid! I'm with you." She tickled his ear lobe with her tongue.

"Tom," the lift boss said to thirty-three-year-old Thomas Daillebaust, "check the load personally. Make sure it won't shift. I hold you responsible." Daillebaust was the tall, Mohawk, Shift Two boss who came to Slingshot by way of his Manhattan heritage—his great-great…great grandfather Louis Joseph Dybo a.k.a. Chief Great Fire.

"Okay, lads. You heard the man." Daillebaust picked up the first load himself.

The lift boss turned to Rodney Chrietzberg, Shift Three boss and thirty-five-year-old descendant of New York Mohawk David Chrietzberg, who broke his back in the 1980s after six years as a lead high-iron man. "Critz, you got the next load. I want a real short turnaround when this boxcar comes back. In the meantime, have your crew take a breather inside."

※

Daillebaust and his Shift Two crew took the better part of two hours to complete the load. They loaded the gas cylinders first, so their weight would be at the bottom. The individual plank rolls were almost impossibly light, and they were soft and flexible. This made carrying them a cinch, and stuffing them into the freight capsule an easy job. But the pile was big, and Daillebaust carefully checked nearly every item that his crew loaded. The containers of tools, adhesives, and miscellaneous materials went in the front of the capsule so they could be taken out first, since they would be needed immediately upon arrival.

When Shift Two had completed the loading, the lift boss called Alex, who insisted on being present for each of the initial launches—until he was certain that the operation was ready to run smoothly without his personal supervision. Bruce arrived with Alex. Lori showed up shortly thereafter with Margo and Klaus, walking between them, arms linked. She was chatting animatedly about having to change her blouse because, "Would you believe it…a bird like that one there," she pointed out a lesser frigatebird perching at the roof's edge, "dive-bombed me and dropped a load of crap right here." She pointed to the top of her right breast. "What a mess…ugh!"

"Were you eating anything?" Margo asked.

"Actually, I was…a granola bar."

Klaus laughed. "The stupid bird wanted your snack," he said. "They're forever attacking other birds with a fresh catch."

"Oh, I've got to see that," Lori said. "Maybe Dex can get it on holo."

On the lift boss' signal, Daillebaust and his crew joined the observers out of line with the boost cable. Then the capsule clamshell cover closed, and the track mechanism moved the capsule into launch position.

"You ready up there, Swamp?" the lift boss said over his dedicated link.

"Great view from up here, but we're getting tired of sitting on our thumbs," Swamp responded. "*Shahsero:ten!*"

"What does that mean?" Lori asked Alex.

"It's Mohawk for *put the lights on*," Alex answered. "Since there's no good way to say *liftoff*, or *launch it*, or *let her rip* in Mohawk, I guess the Mohawks have substituted that phrase."

"*Shahsero:ten*...I like it," Lori said. "*Shahsero:ten*...you got that, Dex?"

He gave her a thumbs-up with a wide grin.

The formal countdown reached zero—and the capsule vanished. The count-up commenced. Five minutes later, on schedule, Swamp announced, "The *kakarénies* has arrived."

Lori looked at Alex. "Freight train," he said. "It's the nearest approximation."

AMELIA EARHART SKYPORT

For over two hours Lars Watson and the other six Shift One members sat on the hardened thirty-centimeter-wide fan elements. Each man had snapped his safety line to one of the elements, and everyone seemed fascinated by what lay below. Watson took one look down and instinctively grabbed the vertical pole. A couple of the guys noticed his action.

"Left your skylegs downstairs, Bob?" Lance Fairbank sneered at him.

"Give him a break," LaFleur said. "Bob doesn't have your sky time, and he's never been this far up."

"No shit, Dick Tracy...and you have?"

Watson breathed deeply, and commenced a bit of damage control. "I got a bit queasy during the freefall," he said. "I'm not quite over it. You looked a bit green yourself when we stopped, Lance."

"Yeah, but look at me now." Fairbank unhooked his safety line and began dancing from one fan element to the other, as far out as he could get, and still keep his 163-centimeter stocky frame upright. "Ring around the rosy—wanna join me, pretty boy?"

"Knock it off!" Swamp ordered. "Lance, buckle yourself up… NOW!" He paused. "Weingard—shuddup and sit down!"

Watson sat straddling a fan element, back against the vertical pole. He tried leaning back, but his helmet interfered so that he had to remain in a slightly leaning-forward position. Fairbank lay across three members so that his helmet protruded downward and his legs swung freely. "Cool way to get paid," he muttered.

LaFleur sat down next to Watson. "Don't mind that jerk," he said. "He means no harm." Fairbank ignored them.

Swamp checked each man's safety line and then joined LeFleur and Watson. "You doing okay, Weingard?" he asked.

"Yah…I guess so."

Watson sat there thinking about what Alex had said before they left: "There are no safety nets." He reviewed what he had experienced for the last two hours. *We're in orbit…but not weightless…if I lost my footing…Ohmygod! I'd fall eighty klicks…* And the light came on for the first time. He was not in orbit. He was sitting on the cap of a tall mushroom, an eighty-kilometer tall mushroom. For the last two hours, he had been thinking they were in orbit like the astronauts in the International Space Station. All his planning for finding a way to float out over the rail and scatter iron filings to gum up the rail—that was all crazy thinking, because he couldn't float up here. *I would fall like the proverbial brick.* For the remaining shift, as Watson worked alongside the others, part of his mind continued mulling over his new understanding. It became pretty clear to him that he wanted to be somewhere else when the rail crashed.

The view from *Amelia Earhart Skyport* was remarkable from several viewpoints. Since they were eighty kilometers above the Earth's surface, there was a distinct disconnect. The sense was not so much that of being on a tall building as being in a stationary aircraft or balloon. Then, because they were so high, most of the Earth's atmosphere was below them. For all practical purposes, they were in a hard vacuum. There was no diffusion of the sun's rays. When they looked down and out, they saw the broad expanse of the Earth completely filling their view, but since they were above most of the atmosphere, the edges of the downward view were fuzzy as the atmosphere thinned, and the Earth's curvature was distinctly visible.

The view upward and away was a black sky filled with stars. When Watson looked directly at the sun, his helmet instantly darkened to protect his eyes from the sun's intensity, and this darkening blotted out the fainter stars, but even with this dimming, the sky still held more stars than he or any other crew member had ever seen before, even on the darkest night miles away from any human activity. The view in any direction was awe-inspiring. The two-hour wait for the freight capsule went by very quickly as the seven men lost themselves in the surrounding view.

Following the short two hours, the freight capsule countdown broadcast to their common link, interrupting their reveries. When the capsule arrived about five minutes later, Swamp announced, "The *kakarénies* has arrived." He shifted to local and began to issue instructions.

"Get the adhesive tools up here, and we'll tie 'em down. Then we'll get a deck under our feet."

Fairbank climbed into the capsule to pass materials to LeFleur, who had secured himself to the ladder. He passed them to Watson, who relayed them to whoever stood behind him. Swamp orchestrated the unloading and placement of materials, making sure nothing was left unsecured. The last thing they wanted was to drop something for an eighty-kilometer plunge to the surface.

An hour later, the capsule was empty and on its way back down the skytower. The first task was to string a safety net below the entire structure. The strands of the net consisted of the same material that went into the stabilizing cables. The individual strands were very thin, and the entire net weighed only a few kilograms. The guys anchored the net along the north end of the fan, and then worked it underneath the elements southward until it covered the entire fan. The net consisted of the main section plus a ring of netting all around the main net that was two meters wide. The crew strung a thin, flexible tube through rings attached to the outside of the net ring all the way round the fan. Then they inflated the tube so that it popped out, extending the net ring into a two-meter wide safety catch entirely around the structure. Within minutes the tube cured, and the safety net was completely in place.

Faster than Watson had realized it could happen, the crew inflated and assembled the planks. They cured rapidly in the broad daylight. As soon as a plank hardened, two men would lay it near its intended position, place a dab of adhesive on the fan elements where it would cross, and then lay a bead of adhesive along its interlocking edge. Then they pressed it into position. The adhesive cured in hard vacuum, so that signaled inspection time for Watson. The crew worked from the center out, inflating the appropriate numbered planks and placing them in a north-south orientation. First, they laid East-One (E-1) on the rail side of the fan. Then they laid West-One (W-1), then E-2, W-2, and so on. Watson moved across the completed flooring, feeling ever more secure as the solid surface grew.

As the surface grew, Watson began to see the rudimentary outline of *Amelia Earhart Skyport*. There was not yet any indication of the extent of the entire station, but the basic floor plan of its heart was already visible: a circular dome with a deck at about the level of the rail itself. A couple of large penetrations through the deck on the rail side marked the future locations of a stairway and a capsule lock. The supporting cables from the vertical pole seemed to pass through everything, and Watson remained unsure how this would be resolved. He had to admit that the whole thing was pretty impressive, and then he reminded himself that his mission was to crash this beast. He jerked his mind away from every external thought and concentrated on his specific inspections. *For now, anyway, I'd better do these right. Wouldn't want things to come crashing down before I am ready.* These thoughts kept Watson focused on the immediate job.

Toward the end of their eight hours, the second freight capsule arrived. They were able to unload it with significantly greater ease, since they could now stow the incoming material anywhere there was a tie-down, so long as they maintained a reasonable balance from side to side and front to back. That was Swamp's job. A half-hour after its arrival, they sent it back empty, and twenty minutes after that, the passenger capsule arrived with Shift Two ready to take over the task of constructing the world's first stationary skyport.

※

Daillebaust and his crew wasted little time admiring the expansive view as the terminator dividing day from night appeared on the horizon a thousand kilometers to the east. By the time it passed below them thirty-six minutes later, they had organized themselves and set to their assigned tasks. In another thirty-six minutes, while they were still experiencing the residual glow of the Sun below the western horizon, the Earth below them as far as they could see was in total darkness, except for brilliantly lit Baker Island directly below them, and the nearly as brilliantly lit nearby Howland Island to the north. The crew installed bright floodlights, but without the sun, there was no way to cure inflated pieces. The planners had allowed for this, however. Included with the second freight load was a portable curing chamber constructed of a flexible Kevlar and synthetic polymer complex. Inflated elements were placed inside this chamber, where they were bombarded with the same spectrum of UV that the sun supplied.

The construction task ahead of Shift Two was to extend vertical members downward to form a matrix that would support a lower deck. With the safety net in place, this task was made relatively easy. Exercising extra caution, Daillebaust required each man who climbed under the deck onto the net to have a safety line that was belayed topside by another crew member. Entry to the underside was through one of the openings Shift One had created during their laying of the deck planks. Although the adhesive was designed to form a bond stronger than the material it was bonding, the designers took no chances. In addition to the adhesive bonds holding the vertical members to the deck, each piece also was physically fastened with three complex polymer bolts at right angles to each other.

By the end of the shift, the guys had installed all the vertical members and interconnected them with horizontal pieces. The subdeck was ready to be surfaced with planking, just like the main deck. Before the arrival of Shift three, Baker sent up another freight load. Daillebaust had his guys secure the material on the main deck near the curing chamber. He'd sent the capsule down, but had not yet heard the countdown for the passenger capsule with Chrietzberg and his Shift Three boys.

Suddenly, without warning, Kelly Seidell, a 183-centimeter tall well-muscled Texan in his late twenties, screamed and dropped face down to the deck. Daillebaust kneeled beside him and gently checked him through his suit. As Daillebaust's hands reached Seidell's right leg, Seidell screamed again. He examined the thigh area more closely, and discovered a jagged hole that the self-sealing suit material had patched.

"You two, give me a hand!" Daillebaust said to the nearest guys. "Help me roll him over." They did, and another patched hole appeared on the front of Seidell's leg opposite the other one, but higher up on the thigh.

Seidell moaned. "My leg's broken," he gasped through pain. "It fucking hurts like hell!"

Daillebaust carefully examined the deck near where Seidell had been standing. There it was—a half-centimeter hole punched right through the deck.

"Baker—we've got a medical emergency up here. Seidell has been hit by something, a small meteorite or piece of space debris, I think. It punched through his upper thigh and then right through the deck. His leg's broke. We need to get him down immediately."

CHAPTER TWENTY-ONE

BAKER ISLAND—SOCKET COMPOUND

Daillebaust's emergency call caused a flurry of activity at Baker Compound. Slingshot had no resident physician, but several divers at both complexes were trained emergency medical technicians. Although the hour was late, Alex called Emmett Bihm, who was asleep in his quarters at the Western Complex. "Bimmy, we got a medical emergency on *Earhart Skyport*."

"Gimme a sec, Boss. Gotta shake the cobwebs." The open link remained quiet for about fifteen seconds. "Okay…shoot."

Alex told Bihm what he knew about Seidell's condition, which was precious little. "Who's your closest EMT?"

"You're in luck, Boss. I didn't tell you, 'cause you've been pretty busy, but I sent Apryl to Baker with Bruce. I figured with everything going on there, somethin' might happen…"

"That's really good news, Bimmy, I…"

"I heard you needed me, Boss."

Alex turned at the sound of the little girl voice behind him to see the 166-centimeter well-toned blonde, pixie hair a bit tousled, no make-up, medical kit slung across her chest resting on her right hip. Despite the late hour and obviously having just awakened, Apryl Searson—diver and EMT—managed to convey both professional readiness and coquettish playfulness.

"She's here, Bimmy…bye.… "

"You're a sight for sore eyes, Apryl," Alex said with a relieved grin, playing to her coquettish side. Then he assumed his professional face and briefed her on what had apparently happened. "Tom believes his femur is shattered. He's in tremendous pain…"

Apryl interrupted. "Has the next shift capsule left yet?"

"No," Alex said, "but you can't go with them—no training, no suit."

"C'mon, boss," she said, putting as much force into her girly voice as she could, "I'm a diver. I wear suits and breathe gas all the time."

Alex remained adamant. "No, Apryl. I'll grant you can handle it in an emergency, but by the time we get you up there, he could be down here, where you can get his suit off and really check him out. You can't do much more up there than what Tom's already done."

Klaus stepped into the conversation. "Alex, we're going to slow down the deceleration rate during descent to minimize the gees." Alex looked at him with concern. "I know, we haven't done it yet, but it's built into the system. You know the system's worked flawlessly thus far. It's a small risk…and the alternative is messing up that shattered bone even more."

"Get yourself set up in the empty room next to Outfitting, Apryl." He turned to Klaus. "Klaus, get Shift Three on its way and take charge of the descent. Leave a couple of guys up there if necessary. Margo…" He turned to look for her and saw her chatting with Lori, who somehow had gotten word of the problem, and had appeared coiffed, made up, and camera-ready beautiful. Dex was recording their conversation. Alex shook his head in amazement. "Margo, can you set up evacuation…"

"Already done, Alex. Lori was just explaining that she was still awake, collecting her notes when she heard about the accident. She got her boss out of bed. Her network has an executive jet in Samoa doing a story on the *Aku Aku*. It's already airborne with an orthopedic surgeon and small operating team. They expect to touch down at Howland in about an hour." Margo hugged Lori.

"You guys are way ahead of me." Alex grinned at them with obvious relief. "Apryl, cancel the empty room. By the time Seidell arrives, the Chinook will be waiting on the tarmac." He tossed an unspoken question at Margo. She nodded, letting him know the Chinook was on its way. "You get Seidell on that chopper and stabilize him on your way to Howland. And Apryl," he added, "if possible, strip the suit off. Don't cut it unless you have to. They're expensive as hell." Then he held his hand up. "But, it's your call… Seidell's interest comes first!"

AMELIA EARHART SKYPORT

The capsule arrived at *Earhart Skyport* a few minutes after Daillebaust made his emergency report to Baker. Shift Three members exited into the lower deck scaffolding and climbed to the main deck through the stairwell opening. The lower and upper decks were brilliantly lit. With virtually no atmosphere, the stars still were brightly visible overhead, but the guys were focused on loading Seidell into the capsule and getting him back on the ground. Time enough later for sightseeing.

"Critz, did you bring any morphine?" Daillebaust asked Shift Three boss Rodney Chrietzberg, as he loosened the makeshift tourniquet on the wounded man's leg.

"Nope. Doc says she wants to examine him before dulling his symptoms. Seemed to know what she was talking about." He paused. "That doc's one good-looking chick, you know."

While they were awaiting the capsule, they splinted Seidell's leg with two plank sections. Meanwhile, the crew brainstormed a way to keep Seidell's leg extended and the patient flat during the return. They came up with a hammock-like stretcher made from a piece of webbing cut out of the circular safety net and several pieces of plank and tubing kluged together with adhesive. They attached the frame legs to the capsule deck with adhesive, taking up five places. That left room for only three guys to return with Seidell. Daillebaust chose Seidell's best friend, fellow Texan Mathew Munns, Chicagoan Guy Roth, and himself.

Fifteen minutes after Shift Three arrived, the capsule dropped down the cable and disappeared in the night sky.

BAKER ISLAND—BAKER SOCKET

The trip down took only eighteen minutes, but to Apryl it seemed like forever. She had been so excited when Bihm sent her to Baker with Bruce. It gave her the unexpected opportunity to see the historic events first hand. She saw her EMT role as a means to an end, and was especially grateful that she had taken the training before coming to *Slingshot*. As she waited anxiously for the capsule to arrive, she nervously assessed her bag of tricks. She knew how to treat someone with the bends and how to stabilize an embolism; she could run any recompression treatment table. She knew how to handle an open wound, to set a bone, intubate; she was expert at CPR—but nowhere in anything she had ever learned did she receive instructions on how to handle a patient whose leg had been shattered by a meteorite.

Alex stepped up to her and placed a comforting arm around her shoulders. "You'll do fine, Apryl," he told her with a squeeze. "You're here because you're good at what you do. Just make your best call and then do what you were trained to do."

Apryl shuddered and looked up at Alex. For the moment at least, her coquettishness was replaced by serious intent. "Thanks, Boss," she said, and kissed him on the cheek.

The capsule arrived and settled to the track gently. As soon as it stopped moving, the gull-wing hatches opened, and the nose cone swung up. Daillebaust, Munns, and Roth quickly manipulated their makeshift hammock out of the capsule and looked at Apryl. "Where do we put him, Doc?" he said to Apryl as he twisted and removed Seidell's helmet.

"Wait a second," she told them as she loosened the tourniquet, causing Seidell to yelp with pain. She quickly checked his heart and breathing.

"I can't take much more of this pain," Seidell told her through clenched teeth.

She smiled warmly at him. "Take it easy, Big Boy. I'll handle that in just a minute." She turned to Daillebaust. "Put him in the back of the pickup. We'll drive him to the tarmac, where a chopper is waiting. I need you guys to come with me. You can strip your gear in the chopper." She waved Alex over. "Alex, how tight is that suit?"

Alex briefly explained the suit, and how the inner layer formed a skin-tight membrane to protect the body from the zero-pressure environment at altitude.

Apryl breathed a sigh of relief. Seidell was in shock, but the pressure of the inner suit against his legs and abdomen had acted like a MAST suit, stabilizing his system. If she could just get some fluid into him to keep him stable, she could strip the suit and address his wound. "Is the surgeon here yet?" she asked Alex.

"He just landed." Alex gave her a direct link. She jumped into the back of the pickup with the three high-iron men and the patient, and they sped off to the chopper.

"Doctor," Apryl said as soon as she established a link, "here's the situation." She briefed him quickly.

"Have you done an IV?" the doctor asked.

"Not yet. I'm stripping the suit to his waist right now. The suit's like a MAST. Gotta keep the bottom on for now."

"Good," the doctor said.

"Get his arms free," Apryl said to the men beside her. "I need to insert an IV." She reached into her bag. "My name's Apryl—with a 'y'."

Daillebaust laughed. "No shit! Really?"

"Yep."

"I can deal with that," he told her as he rolled the suit sleeve down Seidell's arm.

"Maybe, but not now," she said to him with a wink. "Let's save this big boy's ass."

"Yes, Ma'am," Daillebaust said, looking at his two companions. They nudged him, and a few seconds later, both Seidell's arms were free.

Apryl carefully inserted an IV and attached it to a bag of TKO saline. She handed it to Munns. "Here—what's your name?"

"Munns…Matt Munns."

"Okay, Matt. Hold this a half meter above his body at all times. When it starts getting empty, let me know." She handed him the TKO bag.

"Yes'm."

Then she injected a dose of morphine into the feed tube, and Seidell sighed in obvious relief as it took effect. "You're an angel from heaven," he said dreamily.

They arrived next to the Chinook, where Daillebaust and Roth carried the stretcher inside with Munn walking alongside, holding the IV bag.

"I have a TKO in and gave him a minimum dose of morphine, and we're boarding the chopper now," Apryl told the doctor. "His vitals are strong, but his BP's low, and he's still in shock."

"How long is the trip?" the doctor asked.

"About fifteen minutes or so."

"Okay…Monitor him closely. Keep the suit on, and increase the IV rate to as much as he will take. We don't know what's going on inside that suit leg."

Alex's simulacrum appeared in Apryl's holodisplay. "How's it going, Apryl?" he asked her.

"Fine, Boss. I'm working with the doctor. We'll be there in ten minutes. You want me to stay with him?"

"Stay until you're sure the doctor and his team have control of the situation. Then you and the high-iron guys come back. They need to rest up for the next shift, and we need you here." As an afterthought, he added, "Tell them to try and save the suit…but Seidell comes first!"

HOWLAND ISLAND—OPERATIONS CENTER

Apryl glanced through a side port of the parked Chinook to see a sleek executive jet roll to a stop at the end of the runway. A pickup pulled alongside the jet, and several men unloaded equipment from the jet into the bed of the pickup. What appeared to be two men and a woman disembarked from the plane and crowded into the cab, and the workers jumped into the bed with the equipment. While Apryl supervised offloading Seidell and moving him onto a table in the Admin Building first-floor conference room, the pickup arrived at the main entrance, and the guys in the bed jumped to the ground and brought the equipment into the conference room. The woman, who turned out to be an operating room nurse, supervised the placement of the operation light and portable stands. They draped another table with sterile cloths, and the nurse moved the surgical instruments into their proper positions.

When everything was ready, Daillebaust and Munns, assisted by two of the pickup crew, hoisted Seidell onto the sterile table cover while the two doctors sanitized their hands and donned surgical gloves.

It took less than a minute for the doctors to determine that the suit leg had to be cut off. Apryl asked them to strip off the other leg, and she had Daillebaust return the suit to the Chinook. Then she stood back and observed the team go to work on Seidell's shattered femur.

"I gotta get back," she told the nurse. "I'll be continuously available by link." She and Munn left to join Daillebaust, and five minutes later, they were airborne.

AMELIA EARHART SKYPORT

After the capsule left with Seidell, Chrietzberg set his crew to installing plank flooring on the lower deck. The process was much the same as with the main deck. They unrolled, inflated, and cured each plank using the curing chamber since it was still nighttime, and then installed them from the center out. When the first layer was completed, they installed spacers and then the top layer on them. An empty capsule arrived about an hour after Seidell left to take the three remaining Shift Two members back to Baker, and about halfway through the shift, another freight capsule arrived with additional materials.

"Hey Critz," one of the guys asked, "who's next. I mean, I thought the chances of being hit by a meteorite were next to zero."

"Not zero," someone piped up, "obviously!"

"Or maybe space debris," someone else quipped.

"Tell me, guys, how many astronauts on the ISS ever got hit?" Chrietzberg asked casually. That seemed to put the question to bed for the time being.

About three hours from shift's end, the eastern sky took on a beautiful glow, and then the terminator passed the horizon. Although it was not a sharply defined line as on the Moon, it still clearly divided night from day. None of the crewmembers had ever seen anything like it, and Chrietzberg called a fifteen-minute break so the men could take it in. With the arrival of dawn, for the first time, the Shift Three crew got a glimpse of the incredible view from the skyport—a thousand kilometers out in all directions. Directly below, the view was sharply defined, but as their gaze lifted, things became increasingly hazy, until sky and horizon merged into a purple-blue haze that rapidly deepened into the velvet black of space, filled with a panoply of stars, hard and bright as diamonds. During the break, the entire crew went to the upper deck, where the view was unobstructed except for straight down. Fifteen minutes passed all too soon.

"C'mon guys…turn to. We got a station to build."

✺

Over the next several days, *Amelia Earhart Skyport* took shape. The structure itself was completed. It consisted of a reflective Mylar-covered geodesic dome seventeen meters high and thirty meters wide sitting on a rim wall of three meters. Large windows circled the dome at eye height. Flush with the dome surface, they consisted of a transparent polymer similar to the *Wampus* and *Wallus* spheres. This polymer had two polarizing layers whose alignment was governed by the intensity of incoming radiation—the brighter the light, the more polarized the windows. Several of the dome's geodesic sections had been replaced with this transparent polymer. They were positioned to give the dome interior eye appeal and supply a sense of personal presence with the stars. The lower deck extended another three meters below the base of the wall, and had its own circle of polymer windows and an arc-shaped transparent floor section looking down the skytower to Baker Island, eighty kilometers below. The smooth underside was broken by heat radiating fins arranged radially around the skytower that entered at the center of the lower deck. Several antenna dishes protruded from the bottom as well. The entire thing looked like a top-heavy mushroom.

The rail exited the dome to the east through a ten-meter long horizontal half-tube that looked like a cylinder cut in half lengthwise. Capsules remained outside the dome's pressure boundary. A capsule arrived into a wide vertical tube at the skytower terminus, where it was picked up by a gantry that pivoted it to horizontal and moved it forward to the capsule dock. The capsule dock was a separate chamber with an airlock into the main deck. It had a scoop-shaped rectangular opening sufficiently large to accommodate a passenger capsule's gull-wing doors or the clamshell of a freight capsule. An overhead sliding door on the dome side of the dock sealed the opening with the dome pressure. When an arriving capsule stopped at the dock, hydraulic rams pressed it against the seal surrounding the opening, and the door slid up.

In normal operation, transferring a capsule to the rail was automatic. Nevertheless, a six-person human crew occupied the skyport at all times to handle potential problems and maintain the mechanical systems. To minimize the need to interrupt the upward transport of people and equipment, the skyport contained living accommodations for the six, consisting of sleeping quarters and a food preparation area. The crew stood eight-hour watches where two individuals conducted maintenance and were otherwise available. In practice, typically, two people slept at any given time, while the other four were up and about. The crews rotated every seven days. When Earthside, the skyport crews operated and maintained the socket facilities.

The skyport held a supply of launch pouches and kick thrusters that had to be attached to each capsule before launch. Since Slingshot was theoretically able to launch a capsule every ten minutes, a twenty-four-hour supply consisting of 144 pairs of launch pouches and kick thrusters had to be on hand at all times. In practice, after initial onload, replacements would be brought up at least twice daily. The empty capsules would be sent along the rail to *Noonan Skyport* for return to the surface.

The underlying concept of *Earhart Skyport* was utilitarian. Inevitably it would become a tourist attraction, however, a destination of sorts for people who wanted the experience of riding into space, but had nowhere to go after reaching the skyport. To accommodate these visitors, the designers went beyond bare utilitarian functionality. The windows and transparent geodesic sections were part of this plan, of course. But beyond the view to the outside, the dome's interior received a designer's touch. Overall, it had more the appearance and feel of a luxury cruise ship than a navy man-o-war. The public areas were designed with this in mind. A visitor could arrive and disembark into *Earhart Skyport*, to spend a few comfortable hours admiring the view, taking in a meal with fellow visitors, and watching the action. Those with more adventurous spirits could hitch a ride to *Noonan Skyport* by riding the rail on an orbital transfer capsule. At *Noonan*, they could stay for a while or return immediately to the surface on the *Jarvis Skytower*. Another option was to board a capsule that would launch into an eccentric orbit for return through the atmosphere to a controlled landing on the Howland Island runway following one revolution of the Earth. *Earhart Skyport* visitors not wishing to ride the rail could take one of the three scheduled capsules that returned daily to Baker instead of making the round trip to Jarvis. Although the skyports accommodated visitors with style, their primary purpose remained as way stations for space-bound capsules.

BAKER ISLAND—BAKER SOCKET

Under a warm early morning azure sky flecked with clouds, Bruce climbed into the third bay of the passenger capsule at the Baker loading ramp. He was excited beyond words, barely noticing when Lori slid into the seat across the aisle beside him. He didn't notice the skreeghing birds looking for breakfast or the thunderheads along the eastern horizon signaling a change in the weather. He was going to space, and that was all that mattered! Unlike the capsules used during the early construction phases, this capsule was designed to accommodate people in street dress. There still were only eight seats, but they looked more like first-class airline seats. Their surfaces consisted of cushioning memory foam—even for the legs, although they still contained restraint harnesses like the earlier capsules.

Dex Lao remained on the boarding ramp, recording the event. Alex entered the first bay, leaving the left seat for Dex. Klaus and Margo took the second bay. Apryl Searson shared the last bay with Carmine Endsley. Alex had explained to Bruce that he was bringing Apryl along as a reward for her efforts to save Kelly Seidell's life, and Carey for his contributions in saving Slingshot. Bruce was pleased. He liked Apryl because she was… well, she was *Apryl*, simultaneously delightfully coquettish and fearsomely competent. Bruce liked Carey because he was awed by his computer expertise, and especially pleased that he had seen the light and shifted sides.

Lori reached behind her and took Apryl's hand. "Apryl, I'm thrilled to see you again. I was so impressed by how you handled that Texas cowboy's problem." She squeezed Apryl's hand warmly. "We'll have to chat about it in the skyport. It will be a wonderful backdrop for an on-camera interview." Lori winked at Bruce and said, "How's my favorite journalist?" Bruce blushed. "Carey," she said, smiling warmly at him. Carey managed to stammer a greeting, but he didn't make eye contact with her.

"Everyone comfortable?" Alex asked over the general call circuit as he stood looking back over the passengers. Heads nodded; Bruce gave him a thumbs-up. Alex waved Dex to his seat. "Anybody nervous?" A nervous chuckle passed through the capsule. "Don't worry. We've been doing this a dozen or more times a day for a couple months now. The geniuses who designed the skytower and the engineers who built it did their jobs very well." A patter of applause rippled through the capsule. "Margo…Klaus… thanks!" More applause. Lori reached across the aisle and squeezed Bruce's hand, giving him a warm smile. She mouthed the words, *I'm proud of you too!* He blushed again, enjoying the warmth that spread through him every time she touched him.

※

Alex caught Margo's eye. After all this time, he still was struck by her unique combination of beauty, brains, and competence. He was beginning to believe that he loved her. Margo smiled warmly at him and then turned to say something to Klaus. Alex couldn't hear what she said and found himself, surprisingly, feeling a hint of jealousy. He chided himself for feeling that way and grinned at his friend and colleague Klaus. *I know she's attracted to me…I know it! But I'm no idiot. She's got a thing for Klaus too.* Alex pulled his gaze away from them and swept the other passengers. *Easy Lad…focus!* Lori caught his eye and stuck the tip of her tongue out at him, giving him a wink. *Damn that woman!* He shook his head with a grin.

"You've all heard the prelaunch time and again. You probably know it by heart, so I'm not going to repeat it. Just stay calm and keep your barf bag handy…just in case." He signaled, and the gull wings closed, hermetically sealing them in their isolated environment.

※

If only Dad could see me now, Bruce thought as the capsule moved forward and pitched to vertical. Oh man…. Bruce listened to the clanking sounds as the capsule uncoupled from the loading track and positioned itself for the lift.

"Normal sounds, folks," Alex told the passengers on the speaker.

"Hey, Apryl…you cool?" Bruce said over his shoulder.

"Yeah, Bruce," followed by a nervous giggle.

"Didja ever think, Carey?" Bruce turned his head, but the seat blocked his view of the seat behind him.

"Hey man, like this beats the *Avenger* all to hell!" Carey squeaked back.

The short countdown commenced. ...*three...two...one...*

Bruce gripped his armrests as he felt himself being pressed into the memory foam. He had not really known what to expect, but this wasn't that bad. As the pressure increased, he flashed back to waking up one night on *Green Avenger* to find a naked Carmina sitting on him, riding him for all she was worth. He grinned at the memory and blushed a bit. The pressure was about the same. He turned his head left and looked at Lori. Her head was turned so she could watch him. She mouthed the words *Sweet Boy* and winked at him. The capsule was in total silence except for a faint high-pitched whine of the gyro and the hissing of air on the capsule's skin. A sharp shudder announced passing the sound barrier, followed by a gradual decrease in the pressure on his chest. A few seconds later, the capsule shuddered again as it shifted from boost to lift cable. As normal weight returned, Bruce turned his head to look out the window. At eye level in the distance, he saw several towering thunderheads. They dropped below his vision as he watched, and he sensed rather than heard the hissing fade away. Bruce's senses reported no movement at all. He felt exactly the way he felt just before the lift—except that then his eyes took in the launch ramp and buildings, whereas now he saw nothing but darkening sky.

Bruce enjoyed the view while he replayed Carmina in his mind for a minute or so. A clang interrupted his reverie followed immediately by his senses throwing him into a panic. *Ohmygod...the capsule is falling! We're gonna die! Ohmygod!*

"It's okay, folks," Alex announced. "We'll be in freefall just for another minute or so. Use the bag if you must."

Bruce swallowed hard and forced his mind to assume control. Just like those strange pictures where you see one thing clearly, and then suddenly you see something completely different, a serene sense of floating replaced Bruce's sense of falling. He was a bit startled by the abrupt change and looked over at Lori. Her eyes were closed, her blond hair floating like a halo around her head. *Just like an angel*, he thought, as weight gradually returned and her hair settled down against her pretty face. The return of full weight, followed by a clank, announced their arrival at the *Earhart Skyport*. Lori turned her head, opened her eyes, and smiled at Bruce.

"Was that cool, or what?" she said dreamily, brushing golden strands away from her eyes. Bruce answered with another blush.

CHAPTER TWENTY-TWO

AMELIA EARHART SKYPORT

As the gantry grabbed the capsule with a clank, Margo loosened her harness. She turned to Klaus as the capsule pivoted to horizontal and glided to the lock. "This is pretty slick," she said, as the hydraulic rams pressed the capsule to the seal and the gull wings opened. "We did pretty good, don't you think?"

Dex Lao disembarked first to record the event. He turned his holocams on Lori, who somehow had managed to make herself camera-ready in the few minutes since their arrival at *Earhart Skyport*. She stepped out of the capsule glowing with excitement. She was wearing sky blue silk pants that partially covered her heeled slippers, and a light blue silk blouse that was modest in cut but revealing in the way it showed her lightly restrained figure.

Margo looked at Lori in amazement and let her thoughts briefly drift to their night together at the Eastern Complex. *What a fascinating, many-sided woman*, she thought as Lori went into her on-camera mode.

"Welcome to *Amelia Earhart Skyport*," Lori said to her audience while tossing Margo a quick smile. "We have just arrived at the top of this tower in the sky from Baker Island in the equatorial Pacific." Dex scanned around the main deck and then focused on Alex, who was just exiting the capsule. "With us is Alex Regent, the man who made all this possible. Those of you who have been following these reports know Alex well, but for those of you who are joining us for the first time, Alex is the boss. Slingshot is his baby. He is MIT-educated, and brings to the project a background in underwater and aerospace engineering coupled with a keen eye for solving problems on the run. He is a charismatic leader who is admired and loved by the more than two thousand people who have helped him make this happen." She reached out for Alex's hand and drew him near. "Alex Regent." Alex acknowledged her comments with a nod and a smile, and remained standing near her.

Lori turned to Klaus, whose tanned features towered over her. "Klaus Blumenfeld is one of the most important of those people. Klaus, please tell my viewers about your responsibilities." Klaus briefly explained the hundred-meter-wide OTEC generators and the role they played. He described the linear drivers and their purpose. Before he got too technical, Lori smiled at him and said to her audience, "Dr. Blumenfeld comes to Slingshot by way of Darmstadt Polytechnic in Germany and Caltech in California."

"Klaus…please," he said. She placed Klaus next to Alex.

"And this," Lori swept across the deck toward Margo, "is the engineer responsible for all Slingshot's underwater construction, except for Dr. Blumenfeld's…uh…Klaus' OTEC generators and linear drivers. I give you Margo Jackson, by way of Berkeley and Duke University." She paused to let Dex present Margo to her viewing audience. "That's right…I said engineer in charge of…" She reached out and took Margo's hand. "This lady is like no engineer I ever knew!" She smiled warmly at Margo. "Please tell us what it's like to be a beautiful woman doing a job most men would die for—and," she emphasized the word, "doing it better than anyone, according to both Alex and Klaus?" Lori stepped back to give Margo central billing. "What's it like to be one of only a few women in an isolated world with nearly two thousand men?"

Margo dropped her eyes and said demurely, "Really, I hadn't noticed." Lori's eyes twinkled. Margo then told Lori's audience about the nature of her job, and about some of the difficulties she had experienced during the construction. She explained the initial *Green Force* sabotage of the floating rail, and then found herself describing how Alex had handled the problem. "We've got two former *Green Force* members with us right now," she said. Margo reached out her hand. "Come here, Carey." Margo chuckled inwardly at the look of dismay that crossed Carey's face. "This is Carmine Endsley, former *Green Force* member. Now Carey plays a prominent role maintaining the complicated and complex software that controls Slingshot. He was instrumental in preventing *Green Force* from successfully completing their sabotage efforts. We're very proud of him." Margo hoped Lori would not subject him to a live interview, and was relieved when she thanked Carey and turned to Bruce. "Bruce Yoon," Margo said, "is our media liaison. When we found him, he was on a self-imposed deep undercover journalism assignment with *Green Force*. He was preparing to go public with what he knew when we broke their operation. Since then, Bruce has been a valuable member of our team, while simultaneously completing his undergraduate coursework at UCLA by Link." Margo glanced at Lori and smiled slightly.

"So tell me, Bruce," Lori said, "what do you have to say after all that?"

"I never imagined this in my wildest dreams," Bruce said to Lori.

"Do you have anything to tell your fellow students?" Lori asked.

"This is the best project humans have ever embarked on," he said. "It opens a cheap door to space, and it doesn't pollute. The *Green Force* people are way offline on this one. Slingshot exceeds the wildest dreams of anybody who has ever looked at the nighttime stars and longed to visit them." Bruce's passion was palpable throughout the main deck. "This is a new frontier for anyone who wants to go where no one has gone before. We're gonna need people of all stripes, with all kinds of backgrounds, all kinds of training. If you want to be part of the biggest step in human history, find your way to Slingshot!"

"Wow! Thank you, Bruce." Lori smiled at Dex's holocams. "How do I follow that?" she said. "This is Lori Kutcher with Fox Syndicate, signing off from the *Amelia Earhart Skyport* eighty kilometers above Baker Island, at the edge of space."

※

Apryl Searson exited the capsule and wandered across the main deck, eyes darting to and fro, trying to take in everything before she awakened from this fantastic dream. She knew she was good at her job, and she figured being allowed to make this trip was a suitable reward, but being here was way more than what she had expected. It wasn't like being in a space station in orbit—that was certain. But it wasn't like being on top of a very tall building either. More like the world's biggest blimp, she decided. It was obvious to her that the skyport was not yet completed. Details that even she could see were still missing, and if they were still incomplete, she reasoned, many other unseen details must still await completion. She spied the staircase leading to the lower deck. Bruce and Carey joined her as she went below.

Apryl wandered over to the arch-shaped window that looked down the skytower. Bruce joined her while Carey wandered off to explore the deck. At the center of the deck on the other side of the window, the terminus and the enclosed capsule-receiving bay occupied about one-third of the available space. As Apryl and Bruce watched, a freight capsule floated up the cable, noiselessly gliding into the bay. In fact, the only sounds they heard were a faint, high-pitched whine from the gyro that kept the skyport stable, and a faint hiss from the frictionless motion deflector inside the enclosure in front of them. A clank—followed by muffled sounds as the unseen gantry hauled the capsule up and then pivoted it to horizontal.

"Let's check it out," Apryl said excitedly as she ran up the stairs to see what had arrived. Bruce followed her, but Carey stayed below, studying the layout.

Lori and Dex stood outside the capsule dock, waiting for the lock to open. Apryl and Bruce joined them. Apryl understood the concept of locks, since she used them routinely in her diving operations. She pointed to the lock door and explained to Bruce and the others, "That door seals with the pressure in the main deck. The door opens upward. When it is open, if the dock or the lock loses pressure, the door drops with gravity and seals the opening using main deck air pressure. It's closed now as a safety measure even though the dock is at our pressure. You see, if the capsule doesn't seal properly, or if something happens and the dock loses pressure, this keeps us safe. As soon as the capsule is docked, they'll open this door." Apryl peeked through one of the polymer windows into the dock. "The capsule door is opening," she said to no one in particular. Bruce joined her at the window.

"How do you know so much about it?" he asked.

"Just like diving," she said, "'cept the pressure is outside in diving. It's inside here." She bumped Bruce with her hip and gave him a sly wink. "In diving, you keep the pressure out. In space, you keep it in." She turned to find that Lori had recorded the conversation.

"I hope that's okay," Lori asked.

"Sure," Apryl answered, hands on hips, eyes twinkling. "Anything you want."

✺

"Boss," Carey said to Alex, "can I speak with you, like, for a couple minutes?"

"Sure, Carey. What's up?"

"I want to show you something on the lower deck." Carey started down the steps with Alex in tow. He walked over to a free-standing structure that reached from deck to overhead, like a large cylindrical pillar some three meters thick. "This is the kick thruster storage, right?"

Alex nodded.

"They're, like, solid fuel, right?"

Alex nodded again.

"What would happen if one of them, like, exploded?"

"They can't do that," Alex said. "At most, they could burn, but even that's highly remote. And if it should happen, a bottom hatch would automatically open, dumping the contents outside."

"I wouldn't want to be, like, underneath," Carey said. Over time his conversations with Alex had become less strained, and his voice began to sound less squeaky.

"Thought of that," Alex said with a smile. "Each kick thruster carries a drop parachute. We might lose a few, but they won't drop on anyone."

"What if one had a bomb, like, built into it?" Carey said. "Do we know how to, like, check for that?"

Carey watched as Alex considered that possibility. "I don't think we do, Carey. I need to address that." He looked at Carey intently.

"Glad you brought that up, Kid. Keep thinking like that. Discover our weak points." Alex turned to go back to the main deck.

"Thanks, Boss. I'm, like, workin' on it."

BAKER ISLAND—BAKER COMPOUND

Lars Watson, a.k.a. Bob Weingard, sat in his room at the Baker Compound working on his plan to bring down Slingshot. He and the other high-iron guys had two weeks, maybe three, to complete the details on *Earhart Skyport*. Virtually everything was done, except installing the storage racks and belt systems for the launch pouches and kick thrusters. The only thing following that was a shakedown as they completed the remaining cosmetic details. They really didn't need high-iron guys anymore, he knew, but it was more efficient to use the six experienced Baker and Jarvis crews, than to hire new people and bring them up to speed. Other than some good-natured grumbling by the tough high-iron guys about doing wimp's work, the final installations were progressing nicely. Watson had pulled off his deception—no one was the wiser. His unknown benefactor had developed a sabotage plan that he was discussing with Watson over his secure Link.

"The effort to crash Slingshot by attacking the tension cables can't really work anymore," Darius Gotch said. "Now, all the anchors are protected by sonar shields. We would have to cut at least five adjacent cables simultaneously, and that's simply beyond what we can do. It would take a first-world country's navy to accomplish something like that. The linear drivers are equally well protected." He paused. "Slingshot IS vulnerable, though—at the *Earhart Skyport*. The rail, downslope, and skytower all converge at one point beneath the skyport. If we disrupt that point, the entire thing comes down."

"I have about two maybe three more weeks' access to the skyport," Watson reported, having no idea with whom he was speaking. "Then our contract expires, and they'll let us go."

"Don't need access to the skyport for this," Gotch told him. "When does the first load of kick thrusters get sent up?"

"Just before we leave, I think. We've been training with mock-ups. So I think we will do the first load-out."

"Okay. I need some numbers. How long from the moment of freefall until full weight returns?"

"About ninety seconds."

"About or exactly?"

"Within a few seconds. I've never done an actual measurement."

"Time it the next three trips. We need to know exactly how long from freefall to full gravity AND the delay between full gravity and first vertical movement by the gantry." Gotch paused. "Call me back when you have that information."

AMELIA EARHART SKYPORT

Alex motioned Margo and Klaus to join him at a table by one of the windows facing west. Down below, the storm system had moved in and completely blocked their view of Baker, Howland, and the surrounding ocean. From their vantage point, they could see the storm structure, such as it was, but could see no details. From earlier satellite data, they knew it was a large storm that still would be a factor when they returned to Baker later in the afternoon. Alex had no doubts about the integrity and safety of Slingshot or even his base of operations. Everything had been constructed with this kind of storm in mind. Some landscaping might have to be redone, but that was it. He figured the birds would fare much better with than without Slingshot's presence because of the shelter the buildings and structures provided. He suspected the crews manning the Western Complex were having an interesting time, but the semicircular deflector had been designed to remain stable and virtually unmoving under circumstances worse than this storm would bring.

"How're your guys doing at the Western Complex?" Alex asked Klaus.

"Bimmy tells me everything is fine. This one's a pussycat." Klaus chuckled and sipped from a glass of juice, since there was no alcoholic beverage on the skyport—at least not yet.

Alex briefed them on his conversation with Carey. "He's got a point," Alex told them. "How difficult would it be to replace the solid fuel in a kick thruster with explosive?" He looked at each of them. "They're not that different chemically…" His voice trailed off.

"If one were to explode in the locker," Klaus said, "it would make a mess, maybe kill some people if the locker bursts, but it won't bring down the rail."

"What if it explodes just before or at rail insertion—in the launch bay?" Margo asked.

"The rail should withstand it," Alex said. "After all, there's nothing to transmit the blast, so the real problem is debris, and the rail can withstand that."

"Okay," Margo continued, "what about at the motionless friction deflector? Couldn't a blast there sever the rail or at least seriously damage it?"

Klaus pulled out a pen and sketched the intersection node on a napkin. "At the least," he said, "it'll cut the lift cable and probably the skytower." He called up his Link and ran some quick numbers. "Depending on the explosive, if they use a shaped charge near the capsule nose, I think they can pull it off." He leaned back in obvious thought.

"How the hell did we miss this?" Margo asked.

"We're not devious—Carey is," Alex said. "Carey, come and join us." Alex motioned him to their table. Carey sat down, and Alex briefed him on their conversation. "Any more thoughts?"

"I got, like, to the same place you did," Carey said. "It's pretty easy to program a detonator so it'll blow, like, at the right place. Don't know much about explosives, but I could do the program easy 'nuf…" Carey rolled his eyes up and to the left. "Yup…easy 'nuf. No big deal."

Klaus grabbed another napkin. "Look," he sketched, "we protect the locker with a reinforced Kevlar polymer sleeve, and we install breakaway hatches top and bottom. That protects the skyport from a locker explosion. We line the capsule dock and launch tunnel with the same stuff, and we place a blast shield here for good measure." He indicated where the kick thrusters and launch pouches were automatically installed on the capsule. "We design a shield that completely protects the area above the terminus, and wraps around the deflector and as much of the cables as possible, right down to the capsule clamp."

"Time frame?" Alex asked.

"A week for the design, maybe less. Manufacture is a trick. Do we have Mabel take care of it, or do we try something right here—I mean, down there?" He pointed in the general direction of Baker.

BAKER ISLAND—BAKER COMPOUND

Later that afternoon, after the group had descended from their visit to the skyport, Alex reached Mabel in her office. "Mabel, it's Alex. I have an urgent matter to discuss with you." He outlined the problem as he, Margo, and Klaus had developed it, and made sure to mention Carey's role. "The kid brought it to my attention, Mabel. This one slipped by all of us."

"So, what are your recommendations, Alex? Obviously, either I make what you need and ship it, or you come up with one of your miracles and do it on site." She appeared to look directly into Alex's eyes.

"The kick thrusters are due to arrive in about a week and a half, and we will need to send them up right away. I don't think you have the ability to design, manufacture, and get me what I need that quickly." He smiled at her. "Can you do that?"

"You just answered your own question, Alex. You have whatever logistics support you need to build it on-site, including the corporate jet to haul anything you need. Just keep my launch loop safe!" She broke the Link.

※

Lars Watson, a.k.a. Bob Weingard, opened his secure Link and reported to his unknown benefactor: "The average time from zero-gee to one-gee normal at the terminus is ninety seconds plus or minus a second. The average time from one-gee normal to gantry capture is three seconds, plus or minus a second."

He was told that the first incoming shipment of kick thrusters would contain one unit that had propellant replaced with explosive. In appearance, that unit was identical to the others, but could be identified by an embedded chip that would register on his secure Link. His job was to make sure this unit was loaded at or near the front of the capsule. It did not matter if it went up in the first capsule, so long as it was near the front of the capsule in which it traveled.

CHAPTER TWENTY-THREE

BAKER ISLAND—BAKER COMPOUND WORKSHOP

Using his own resources and those available to him through his professional network, Klaus came up with a design and manufacturing technique for each of the required skyport modifications. It turned out that there was no need to install a rigid Kevlar polymer cylinder in the locker, or to invent a way of creating and installing rigid sections in the capsule dock and launch bay. Instead, Klaus came up with a fabric woven from the same material used to construct the planks. Sufficient raw material was available in Seattle, left over from manufacturing the planks. Alex consulted with Mabel, and she promised to have enough fabric woven and flown down within three days. With her connections, she was—for a price—able to find a Seattle source to draw the needed filament, and have it delivered to a local commercial weaving firm that produced the required amount of fabric overnight. Early the third day, the *LLI* corporate jet onloaded the fabric rolls and delivered them to *Amelia Mary Earhart International Airport* late that afternoon. Before sunset the third day, the Chinook deposited the fabric rolls on the Baker tarmac. They were shipped to the skyport following Shift Three's ascent.

Klaus' solution was to attach several layers of fabric to the interior wall of the locker and the appropriate walls of the capsule dock and launch bay with the same adhesive used in the construction of the skyport. Since both the locker and the capsule dock could be exposed to hard vacuum, and the bay was in hard vacuum, the adhesive could cure properly by this exposure and the porous nature of the fabric. Then the locker, dock, and bay would be bathed in UV from a bank of portable lights to cure the fabric to rigidity. The same process would also work at the terminus to protect the underside of the skyport. Unfortunately, the problem of protecting the frictionless motion deflector and the cable bundle at the terminus could not be directly solved with this method.

Using his Link resources to simulate the frictionless motion deflector and cable bundle, Klaus spent the next morning designing a casing that enclosed the frictionless motion deflector at the terminus and enclosed the cable bundle with a snug-fitting sleeve down a meter and a half. He sent Carey to round up as many Styrofoam packing sheets left over from incoming shipments as he could find, and bring them to the workshop. There he had Carey glue the Styrofoam sheets together until he had a block sufficiently large to contain the entire volume of the frictionless motion deflector and the portion of the cable bundle he wanted to protect.

"Carey," Klaus said, "we need to carve that block into this exact shape." He showed Carey the design on his holodisplay. "The best way is to use this focused laser in an x-y-z matrix driven by a computer program." He lifted a small laser unit for Carey to examine. "What do you think? Can you do it?"

"Sure. But why don't you just use a three-D printer to make the sleeve?"

"Wish I could," Klaus said, "but we don't have a large enough three-D printer down here to do that, not to mention the plastic we would require."

"Okay, I, like, get it. So, what kinda, like, output do you need?"

Klaus walked over to a large blueprint-generating table. "This here is, basically, an x-y plotter. The pen is held here," Klaus pointed to the intersection of two right-angle wires, "and x-y signals position both the wires from the edge of the table. A third signal activates the pen. We'll replace the pen with a round platter driven by a step motor. The x-signal will move the plate left and right, like this." He moved his hands across the short width of the table. "The y-signal will step the platter around like this." He twisted his hands over the table. "While you do the programming, I'll make the platter and attach a rod here," he pointed to the middle of the right-hand long edge, "mounted with a step motor, line, and pulley, with a clip for the laser attached to the line. The step motor will take the z signal."

"So…" Carey said, "x moves the block toward or away from the laser, y rotates the block, and z moves the horizontal laser up and down…"

"Simpler than that," Klaus said. "X moves toward and away from the laser as you said, but both y and z simply step—y steps one complete circle around the block at a constant z, repeating at the next z, and so forth down the block." Klaus grinned at him. "Got it?"

"Like…the Styrofoam block moves in and out while turning so the laser, like, burns deeper or shallower all the way around, and then it, like, moves to the next level, and so on?" Klaus nodded.

"Okay…gimme, like, a couple of hours," Carey said.

※

Four hours later, Klaus and Carey stood in the Baker Compound workshop, watching a nearly invisible laser beam burn away excess Styrofoam from the makeshift block.

"Tell me again how you were able to produce that program so quickly," Klaus said.

"I didn't," Carey answered. "I hacked into the company files of, like, one of several dozen firms that makes laser sculpting equipment, and, like, downloaded their control program. Then all I did was, like, adapt that program to this thing here." He pointed to the kluged laser-sculpting rig.

"Cary," Klaus said, shaking his head. "You know we cannot function this way…what you did was illegal."

"Sure, but…"

"I appreciate your enthusiasm, but we're going to have to correct this." Klaus shook his head again. "But first, let's apply what you created." He made a note in his Link to bring the problem to Alex so they could set up either a licensing arrangement or purchase the programs Cary hacked.

A half-hour later, a perfect replica of the enclosed volume of the casing and sleeve stood on the workbench.

"Like, what now?" Carey asked.

"Remember the rail sheath that disintegrated in direct UV?" Klaus asked.

"Sure…"

"We're going to cover this sculpture with the same polymer. Then I'll show you," Klaus said with a smile. The sheath polymer came in liquid form, and Klaus picked up a bucket he had stowed under the workbench, slipped on a pair of latex gloves, and opened the bucket with a screwdriver. From a drawer, he removed a medium-width paintbrush and set to work covering the Styrofoam sculpture with a layer of polymer. Klaus left it to cure, and waved Carey to follow him. "Let's grab a soft drink."

They returned to the workshop a half-hour later.

Klaus unsheathed his knife and cut the now hardened shell in half lengthwise, and removed the Styrofoam. "Carey, over there," he pointed, "in that cabinet is a matrix of UV-emitting LEDs. Please get it and bring it here." Carey pulled it out of the cabinet, looking very much like a net of Christmas tree lights. Klaus pulled on a fresh pair of latex gloves and spread a layer of liquid polymer on the inner surface of both halves. Then he laid the LED matrix into the fresh polymer and positioned the individual LEDs to project above the surface. "We'll attach this temporary shell to the frictionless motion deflector and cover it with a layer of Kevlar polymer fabric," Klaus explained. "Once the Kevlar polymer blast shell has cured, we'll flood the casing with ozone and activate these LEDs. The temporary shell will disintegrate, leaving the blasted shell in place." He clapped Carey's back. "Let's get something to eat and drink, Carey. We've earned it."

An hour later, Klaus introduced Carey to one of his favorite Belgian beers as they washed down thick steaks and fries in the good company of Margo and Alex.

AMELIA EARHART SKYPORT

Lars Watson, a.k.a. Bob Weingard, arrived at *Earhart Skyport* with Shift One a day later. The materials they needed had been shipped aloft during the night, so they could commence the blast modifications immediately. Because they would be working outside the pressure boundary, they had brought their suits with them, but didn't wear them during ascent because Alex didn't want to change out the capsule seats for two shifts. It was easy enough for the crews to carry their suits and don them at the skyport.

Although Alex had no reason to suspect any of his high-iron crew, he decided to compartmentalize the blast-protection tasks to the extent that he did not brief the crews on the extent of the modifications. He briefed the shift bosses on the general nature of the task, since he was convinced that people worked better when they understood what they were doing. But he did not brief each shift on the tasks he had assigned to the other shifts. He didn't make an issue of it either, so the crewmembers were free to talk to each other about their tasks. Consequently, Shift One would install locker, capsule dock, and launch bay protection. Shift Two would install UV lamps to cure Shift One's work. Shift Three would install and cure protection around the frictionless motion deflector, since they had to put the deflector back into normal operation to return to base. Finally, Shift One would remove the curing lamps from their installations. That left about a week to install the racks, belts, and loading mechanisms for the automated storage and installation of the launch pouches and kick thrusters.

Shift boss Mike Swamp handed out the work assignments immediately following their arrival. "Pete, you and Bob got the locker—don't need an inspector, Bob, so you get to see how the other side does things. The rest of you got the capsule dock." He assigned himself to assist LaFleur and Watson in setting up their scaffold suspended from the locker rim where it penetrated the dome. The setup was simple enough. They removed and secured the frangible upper hatch and attached a crosspiece consisting of two plank sections connected at their centers. From this, they suspended a bar designed to hold a roll of Kevlar polymer fabric, and below that, a workbench consisting of two planks. The level of the work platform was controlled by a double-purchase set of pulleys attached to the bar and to the frame at the top, and tied off to cleats on both sides of the bench. They started at the rim, applied adhesive as far down as they could reach, and then laid the fabric on the adhesive, unrolling it from the roll over their heads. After smoothing the fabric, they lowered the bench a couple of meters and completed the next section, repeating the process nine times to cover the length of the cylinder. The locker had a circumference of just under ten meters, so they ended up doing five vertical sections, taking about an hour for each section. It had the advantage of being simple, but the pulleys were entirely manual, which left open the possibility of slipping the line and dumping one or both sides of the bench.

Swamp solved this potential problem by including two safety lines tied off at the rim and the bench. Before lowering the bench, Watson and LaFleur would extend the safety lines by two meters, tie them off, and then lower the bench. Furthermore, each worker had his individual safety line that he used in the same way.

The other four had a much easier task, since they could work in shirtsleeves while inside the landing dock. Once they had completed that task, they suited up and evacuated the air from the dock to vacuum-cure the adhesive. Their remaining tasks consisted of applying the protective fabric on the outside of the dock and in the launch bay, and installing a blast shield over the launch bay deck that extended out to where the launch pouches coupled to the rail. The last proved to be tricky, so that by shift end, they were still installing supporting scaffolding under the shield designed to transmit the force of any blast away from the rail to dissipate in the skyport structure.

As Watson applied adhesive and smoothed fabric in a constantly repetitive series of actions, creating significant blast protection for the structure he intended to bring down, he consoled himself with the knowledge that they were protecting a part of skyport that was not at risk—at least not from him. Despite the repetitiveness of the task and his own ultimate intentions, Watson threw himself into the job at hand. It was so foreign to anything he had ever done, and it challenged him in ways he had not imagined. As he worked the bench up and down inside the locker, from time to time, he stopped at the rim for a brief break and to gaze around the star-filled sky. He and LaFleur sat on the rim, feet dangling over the edge. LaFleur said to him, "Hey Bob…if we were in orbit, we wouldn't need all this crap." He indicated the suspended scaffold. "We'd just tie ourselves off for safety and float our way down the shaft. We'd be done in less than half the time."

Watson grunted and sat there thinking about what LaFleur had just said. *Not floating…* he winced at his earlier misunderstanding… *not weightless…if I lost my footing…*he recalled his sudden insight during his first trip to the skyport…For the remaining shift, as Watson worked alongside LaFleur, part of his mind continued mulling over the facts. And that's when it hit him. When the blast went off next week, the skyport was going to drop straight down—the Baker Compound would be smashed. Watson figured he wanted to be long gone before that happened.

※

Thomas Daillebaust and his Shift Two crew took over at 6:00 pm. They immediately set about curing the Kevlar polymer laid by Shift One. Since he was missing Kelly Seidell, he joined the two guys he assigned to the capsule dock and launch bay. First, they evacuated the capsule dock and exited to the launch bay through the capsule door. They positioned UV flood lamps to cover every part of the bay, and then set up lamps along the gantry rail outside the capsule dock. This part of the skyport actually received a significant amount of sunshine, but Daillebaust wanted to ensure the Kevlar polymer fabric cured to a hard shell. Then they set up lamps inside the dock and exited into the main deck through the airlock. Meanwhile, Mathew Munns and his two guys lowered an array of UV lamps from the brace at the top of the thruster locker so that the entire locker was flooded with ultraviolet light. Following the hardening exposure, they retrieved all the lamps, sealed the capsule dock in its evacuated state, and doffed their suits in preparation for the arrival of Rodney Chrietzberg and Shift Three.

※

Shift Three arrived at midnight, and the capsule departed again almost immediately with Shift Two aboard. Twenty minutes later, a freight capsule arrived with the frictionless motion deflector casing. Chrietzberg's crew was suited up and waiting. Unlike the previous blast-proofing tasks where there was something solid below the guys for each phase, this installation required two guys to be suspended from the terminus with nothing beneath them but eighty kilometers of empty. Chrietzberg decided not to use a safety net, since that would have entailed constructing one back at Baker, and then using more than an hour just to install it. Instead, he put two guys in harnesses, belaying them from the deck in the receiving bay. Their first task was to install the fabric to the overhead area. Although they were surrounded by eerie silence, every time one of them touched the deflector case, he could feel the hum generated by the high-speed cable moving through the deflector. Initially, both workers reacted with a start to the hum they felt rather than heard. Over the next few minutes, however, they got used to the effect and soon forgot it was there. Instead, they concentrated on the awkward overhead position and lack of underfoot leverage that made what appeared a simple task last for nearly two hours.

Installing the deflector mold was relatively easy. Each man held one half. They pressed their halves together around the deflector and temporarily secured them with that old stand-by and builder's friend—duct tape. Then a coat of adhesive, a wrap of fabric, more adhesive, more fabric—until they had built up five layers of Kevlar polymer fabric. Hang the UV flood lamps. Take a break while it cured. Flush the casing with ozone from a suspended bottle. Plug in the LED array.

A few minutes later, the LED matrix fell out of the blast casing as the temporary mold disintegrated from the effects of the ozone and ultraviolet. The task of protecting *Earhart Skyport* from an exploding kick thruster was complete.

Soon thereafter, eight hours had passed, and Shift One was on its way up to replace the tired Shift Three guys. Simultaneously, high over the central Pacific, a transport plane carried four pallets in its belly. Two were stacked with innocuous launch pouches. The other two displayed large, very visible warnings about self-oxidized fire hazards. Hidden deep inside the molded "propellant" of one kick thruster on the fourth pallet, a tiny chip transmitted an undetectable spread-spectrum signal, and waited patiently for a zero-gee signal followed in ninety seconds plus or minus a second by a one-gee signal, followed in three seconds plus or minus a second by a distinct vertical tug.

CHAPTER TWENTY-FOUR

BAKER ISLAND—BAKER SOCKET

The Chinook had already deposited two pallets of launch pouches on the Baker Island tarmac. As it approached for a second time, its twin rotors creating a harmonic sub beat that reverberated all over the small island, thousands of terns, boobies, and frigatebirds took to flight, scattering away from the pounding rhythms. The sky was virtually cloudless following the recent massive storm, and despite hurricane-force winds that gripped the island for an hour or so, the vegetation suffered little damage, and the bird population following the storm was significantly larger than it would have been before the advent of Slingshot. The big helicopter left behind two pallets, each holding twenty-seven kick thrusters.

AMELIA EARHART SKYPORT

Lars Watson, a.k.a. Bob Weingard, was still working with Shift One on *Earhart Skyport* when the pallets arrived. The crew had spent the eight-hour shift installing loading racks in both the launch-pouch and kick-thruster lockers. Although serving a similar purpose, the mechanisms were very different. A launch pouch attached to the undercarriage of a capsule before launch. As its name implies, it served to couple the capsule to the rail magnetically, and launch it into space at a velocity and angle necessary for the capsule to reach its destination. This launch was influenced by three factors—launch velocity, the point of release from the rail, and the relative position of the point of release with respect to the destination. All of this, however, initially depended upon proper launch—the task of the launch pouch. In preparation for launch, the pouch transport belt brought a pouch to the launch bay where a clamp held it from below, suspended over the rail. The gantry moved a launch-ready capsule over the pouch, and the clamp raised the pouch and attached it to the undercarriage of the capsule, where it automatically locked into position.

Simultaneously, a kick thruster was attached to the rear of the capsule, just aft of the stabilizing gyro. The kick thruster served two missions. It was a recently developed solid-state engine that could be throttled up and down, and even be turned off and reignited. The capsule launched itself off the rail into an elliptical transfer orbit by decoupling the pouch from the rail. As it neared its destination, in most cases, it would require minor course and speed adjustments. The kick thruster accomplished these in conjunction with the gyro, which pointed the capsule in the right direction. Where the capsule destination was unobtainable from the initial launch conditions, the kick thruster served to supply the necessary added boost or braking. The thrusters were stored vertically in racks. The transport belt received a thruster from the locker, brought it to the launch bay, and inserted it in a magnetic holder on the back bulkhead of the bay. Just before receiving the pouch, the gantry backed the capsule to the thruster, where it automatically locked to the capsule. When the capsule departed, a protective liner sleeve remained behind, to be recovered for reuse.

At full operational capacity, Slingshot could launch a capsule every ten minutes. In principle, this would require 144 pouches and thrusters each day. The thruster capsules could carry eighteen units, and the pouch capsules could carry twenty-nine units, so thirteen daily lifts were necessary just to keep the capsules boosting into space—at full capacity, of course, which was still down the road a bit.

Shift One had installed the racks and commenced installing the transport belts. Although both lockers were normally evacuated, they needed to be pressurized once or twice daily for reloading. Pressurizing required sealing the back ends where the belts received the devices for transport to the launch bay. This was accomplished by dropping pressure doors over those openings. The pouch opening was a horizontal slot; the thruster opening was round. Shift One ran into a delay with the backside opening mechanisms, and so did not complete its scheduled installation.

Shift Two picked up where Shift One left off. Since Shift One had essentially solved the seal problem, Shift Two completed the seal and belt installations within their scheduled timeframe. Shortly before midnight, the crew finished testing the transport mechanisms, and reported to Alex that they were ready to receive the pouches and thrusters.

BAKER ISLAND—BAKER SOCKET

As soon as Lars Watson returned to Baker, he grabbed Peter LaFleur. "Hey, Pete…I hear the pouches and thrusters have arrived. They're on the tarmac. Let's take a look."

"Good deal, Bob. Then we can grab some supper at the cafeteria."

The two men walked together up the gentle rise to the tarmac a half-kilometer distant. Watson had set his Link responder to vibrate unobtrusively against his wrist. As he and LaFleur approached the pallets, he felt a faint vibration from the Link. His heart skipped a beat—the bomb was on one of the pallets. Watson walked around the four pallets. "Here are the launch pouches," he said, indicating the first pallets to be dropped off. Each pallet held thirty-two pouches.

"Those are two-man carry, for sure," LaFleur said.

"These, too," Watson said, indicating the thrusters.

"They're not as heavy as they look," LaFleur said. "But remember what they said in training? If you hit one hard or drop it, it can't be used."

"Sensitive shit!"

"Makes you appreciate Slingshot," LaFleur said. "No rockets, 'cept these guys."

Watson walked around both thruster pallets and then around each one, surreptitiously holding his Link so its pickup could record both holographically and generally in the electromagnetic spectrum, exactly as his benefactor had instructed him. "Let's grab some chow," he said to the Mohawk.

Their meal completed, Watson excused himself for a head break. In the stall, he called up a holodisplay of the pallets. The display indicated that the bomb was in the bottom layer at a corner of the pallet with the off-color pigmentation, which to Watson's gray-toned vision appeared the darker of the two. Watson carefully examined the display to find an identifying marker so he could easily determine which corner. There—the bottom plank at the corner had a knothole in the edge right under the bomb. It was distinctive, different from the other corners. He left the head as a general call went out for Shift One members to assemble at the loading ramp to load the pouches and thrusters into the freight capsule.

Since the entire shift had been in the cafeteria, they all walked together to the loading ramp. One of the ground crew guys had already transported three of the pallets by forklift from the tarmac to the ramp and was bringing the last one—the off-color one—as Shift One arrived. Watson saw that his pallet would be unloaded first. The crew worked in pairs while Mike Swamp supervised. The crew loaded the nine thrusters from the top layer, carefully stacking them in the capsule racks. Then they loaded the second layer. The capsule lifted several minutes later, and they commenced loading the next capsule with the third layer. Watson grabbed a thruster from the far row, setting the loading pattern. He and LaFleur ended up loading the bomb at the top of the first stack in the capsule. The crew followed that by loading the top layer of the second pallet.

Since the launch rail had not yet been tested, the capsules had to drop down to Baker instead of riding the rail to *Noonan Skyport* for return to Jarvis. So, until the first capsule returned, the second could not be launched. They had expected to launch the second capsule as soon as it was loaded, but Shift Two reported a logistics problem at the skyport. The process of offloading the thrusters from the capsule needed to be rethought, and the method of moving them to the locker needed some adjustments. Tom Daillebaust requested an hour to sort things out.

Watson checked the time. It was going on 10:00 PM. By the time they were ready to accept the next load, it would be close to midnight. He guessed Shift Three would replace Shift Two before the second thruster capsule was sent up. *Good for Tom…bad for Critz*, but Watson never liked that braggart Mohawk anyway. "Gonna get some shuteye," he announced. "We're done for the day." He left for his room in Baker Compound.

BAKER ISLAND—MEYERTON LANDING

"The bomb goes up in about two hours," Watson reported on his secure Link. "I need to be long gone by then."

"Can you get your hands on a powerboat?" Gotch asked.

"No problem."

"Howland Island is sixty-seven kilometers due north from Baker's north shore. Grab that powerboat and some water, and head north…make sure you have four hours of gas. By the time you get to Howland, we'll have tickets waiting for you."

Twenty minutes later, Watson loaded four gas cans into a Zodiac tied to Meyerton Landing wharf under the light of a nearly full moon. A five-knot breeze from the east carried any sound he made out to sea. He dropped a duffel bag with four water bottles and several candy bars on the boards by the big outboard, stepped into the rubber boat, and cast off. One smooth pull on the starting cord brought the engine to life. Watson engaged the gear and steered quietly toward the harbor entrance, crouching low to minimize his silhouette in the bright moonlight. Once outside, he turned north, guided by his Link, and opened his throttle to full.

BAKER ISLAND—ABOARD SKIMMER ONE

Margo sipped a Merlot on the north facing deck of her Baker bungalow. The full moon glistened off the breakers rolling across the reef that blended with the north shore sand. The evening was late, about an hour before midnight, and the light easterly breeze kept her pleasantly cool. The sounds drifting up from the base of the skytower had ceased, the birds had retired for the night, and the only sound came from the breakers as they broke on the reef and bubbled to the sand. As she sipped her wine, the wind shifted around to the north, bringing with it the smells from the reef.

Suddenly she heard something…there it was again…the unmistakable sound of an outboard coming from somewhere beyond the breakers, carried to her on the breeze. She strained to see something, but even in the moonlight, all she saw were the gently rolling waves that broke one after the other on the reef below. The sound faded for a few moments, and then she heard it again. Something was wrong; she was certain of this. She called Alex by Link. "Alex, there's a boat off my beach. It's small, probably a Zodiac. I can hear it, but can't see anything. Something's wrong…I know it. Something's terribly wrong…I can feel it in my bones."

"Get Klaus on this circuit, Margo. Tell him to meet me at *Skimmer One* as fast as he can get there."

"Okay…and I'll join you guys."

Fifteen minutes later, Margo met Alex and Klaus on Meyerton Landing wharf, right behind the spot Watson had left shortly before. They quickly boarded *Skimmer One*, and got underway with Klaus at the helm. In seconds they passed through the harbor entrance and angled northward. Klaus powered up the surface search radar and scanned the water before them. The radar indicated a contact almost immediately, about four kilometers out, and moving fast. Klaus cranked up his thrusters, but kept the skimmer off the cushion, moving through the waves at some forty-five knots. The radar indicated that the boat was doing twenty knots, so they were closing its position at a bit over twenty knots. In a little over five minutes, Klaus switched on his floodlights and directed the radar to aim a powerful spot at the boat's location. Clearly illuminated in the bright beam, a Zodiac raced through the water. It all happened so fast, that Margo nearly missed it.

"Nice job," she said, giving him a hug.

Klaus picked up a hand mike and announced over his broadcast speaker, "You there…you in the Zodiac…heave to immediately!" The Zodiac continued at the same speed. "You cannot escape…heave to…or prepare to be rammed!"

That did it. The zodiac cut its motor to idle and settled lower into the waves. Klaus pulled alongside, so the rubber boat was by the stern well where Alex was waiting with a line and hook. Alex tossed the line across the Zodiac's transom and tied it to a cleat, effectively holding the boat's stern alongside the skimmer. Margo stepped into the stern well and took the line from Alex so he could grab the bow with his hook.

"Okay, get your ass up here."

Reluctantly, Watson complied. As he emerged over the gunwale, Alex looked at him with total astonishment. "Weingard…what the fuck are you doing out here?"

Watson stammered something about an emergency. Alex said to Margo, "This guy is a high-iron inspector from Shift One—Bob Weingard. He arrived with the first group; hangs out with the Mohawk Pete LaFleur." Margo looked at this man, and then it was her turn to register astonishment.

At that moment, Alex received a Link call. "Boss, it's Critz. We're standing by to receive the next load of thrusters."

AMELIA EARHART SKYPORT

The capsule carrying Shift Two left *Earhart Skyport* right on time. Rodney Chrietzberg set his men to preparing for the next load of thrusters. He expected it to arrive in fifteen minutes. He called Alex by Link to let him know.

BAKER ISLAND—BAKER SOCKET

Tom Daillebaust and his crew debarked from the capsule at the socket. They cleared out quickly so the passenger capsule could make way for the thruster capsule. In less than five minutes, the thruster capsule was positioned vertically and ready to lift.

"Let her rip," Alex ordered by Link.

ABOARD *SKIMMER ONE* BETWEEN BAKER AND HOWLAND ISLANDS

Margo heard Alex give the order. "Alex," she said, "this guy's name isn't Weingard…it's Lars Watson." She looked at Alex with alarm. "Something's wrong, Alex. He's running. Something's very wrong!"

Margo turned to watch the lighted capsule race up the skytower into the night sky. At this distance, she figured she might actually be able to follow the capsule's progress for the first few kilometers. Alex and Klaus turned as well. Klaus had a firm grip on Watson's arm, but Watson managed to twist so that he, too, had a view.

Sure enough, Margo could follow the capsule for about fifteen seconds, and then she lost it among the stars. All four stood gazing skyward, but with nothing more to see, Alex and Klaus turned away. Watson continued to gaze intently skyward, even though Margo was certain he could see no more than she. Her sense of unease returned. She reached out to grab Alex's shoulder, to tell him about her unease. As she did, a brilliant flash in the sky high over Baker caught her attention.

"You goddamn sonofabitch!" Alex shouted at Watson, and decked him on the spot.

CHAPTER TWENTY-FIVE

AMELIA EARHART SKYPORT

Eighty kilometers over Baker Island, Rodney Chrietzberg and his team had just checked the status of the lockers. Since they had a few minutes to kill, six of them stood at the panoramic windows admiring the nighttime sky dominated by the full moon. The Earth beneath them was a huge black abyss that filled nearly half their vision. Chrietzberg stood at the terminus window, looking down the skytower. Far below, he could clearly see the brightly illuminated Baker Compound in stark contrast to the black ocean. The full moon occasionally reflected off an ocean wave so that, from time to time, a flash of light reached his eyes from the featureless ocean surface, sparkling like a diamond. As he watched, he began to make out the light atop the rising capsule. After a minute or so, the capsule was clearly visible in the bright moonlight. It shed velocity as it floated up the skytower until it was just meters below him.

"Capsule's here, boys," he said, as it slipped into the receiving bay. The clank as it clamped to the skytower broke the midnight quiet.

A second later, a blinding flash saturated the main deck. Chrietzberg lurched back as a deafening explosion ripped through the receiving bay. The skyport tipped steeply and ominously northward, as the gyros screamed to return it to stability. In a heartbeat, all seven men found themselves at the bottom of a makeshift canyon, wedged between the angled deck and the bulkhead, being pelted with anything that was loose. One of the men screamed in pain…all the men were shouting. The deck lurched further, answered by an even higher pitch from the overworked gyros. Then it settled back a bit, lurched again, and settled back a bit more. A very loud snapping sound penetrated the deck from the receiving bay, followed by a short vertical excursion and a major lurch that threatened to turn the skyport upside down. The gyros screamed even louder, generating a wavering pulse as their rotational frequencies moved in and out of synchronization. Then slowly, with small jerks accompanied by lots of creaks and rumbles, degree-by-degree the skyport righted itself. Chrietzberg climbed to his feet, surveying the situation as he felt the skyport yaw ominously several times, and slowly pitch almost as if it were riding at an angle over a series of ocean breakers. Just as he started to feel queasy from the unaccustomed motion, the gyros took charge and dampened the wayward skyport into submission, and all motion stopped.

Chrietzberg continued his survey, listening carefully for the sound of escaping air. The skyport was silent as a tomb. The deck appeared level to his trained eye, and except for the mess on the north side of the main deck, things seemed normal again. "Who's hurt?" he asked.

"It's Pigman," somebody said. "He broke his arm, but the skin's not broken."

One of the Shift Three guys had replaced Kelly Seidell in Shift Two, and Rudolph Pigman, the capsule attendant for the first capsule lift, had taken his place in Shift Three. Chrietzberg examined his broken arm. "You'll live, Pigman," he said to the stocky Boise native. To the others, he said, "Splint that arm and give him some pain killer from the medical kit. We got a problem to solve."

BAKER ISLAND—BAKER SOCKET

Shift Two arrived groundside and headed to the cafeteria almost immediately. Within a few minutes their capsule was shunted aside and the loaded freight capsule was readied for lift. Shortly thereafter, it disappeared up the cable. The process had become so routine that nobody remained around to watch the process. Five minutes later as the capsule arrived at the terminus, a blinding white light bathed the compound, and all hell broke loose.

Within a few short seconds, cable moving at 400 meters per second crumpled onto the socket, slicing through the skytower cable as it dissipated its massive kinetic energy in a few microseconds. The immediate structures shattered abruptly, torn asunder by the writhing, snaking cable as it fell from the sky. Anybody still in the launch facility died instantly—too fast to feel any pain or know what had happened. Once the falling cable had dissipated its kinetic energy, it piled up where it fell.

Eighty kilometers up in the dark sky, the frictionless motion deflector separated from the skyport several seconds after the explosion. As it commenced plummeting to Earth, twisting and turning as it fell, the kinetic energy of the lift cable still traveling at 400 meters per second jammed the deflector, shattering it. Instantly the lift cable separated, and the upward-moving end whipped off into the sky, trailing the rest of the cable after it in a generally northern direction. The heavy falling deflector seventy-two kilometers below flattened the lift cable's arc as it sped north.

Two minutes and several seconds after the explosion, the double motion deflectors and gyro from the eight-kilometer level crashed into the tangled twenty-four kilometers of cable piled in the center of Baker Island. Forty-eight kilometers of lift cable had already streamed through the lower deflector, arcing into the night sky. When the deflector bundle smashed into the cable pile at nearly 200 kilometers per hour, the gyro shattered, splitting it apart, separating the two deflectors. The lift cable deflector jammed. This caused the skyborne lift cable, which still was traveling at nearly the speed of sound, to snatch the deflector into the sky, dragging both ends of the skytower cable with it, but leaving the boost cable behind. Since the streaming lift cable was moving very much faster than the falling skytower, the falling skytower simply followed the lift cable skyward in a flat arc, no longer accumulating on Baker. The eight kilometers that were already piled atop the socket also followed the lift cable into the sky, scattering the confused coils of boost cable. The increasing energy required to extract the skytower from the twisted pile sucked the remaining kinetic energy from the streaming lift cable. It slowed significantly and began to arc toward the sea. In less than fifteen minutes, the dead weight of the trailing skytower cable caused it to hit the surface starting at Baker Island, splashing into the sea with tremendous force as it followed the lift cable track northward. The trailing deflector smashed into the water just off Howland Island's southern reef, stopping all forward motion. With its forward momentum dissipated, the lift cable snapped into the air like the end of a bullwhip, emitting a sonic crack that could be heard for miles, and dropped across the length of Howland, nearly bisecting its runway, but causing relatively little actual damage.

ABOARD *SKIMMER ONE* BETWEEN BAKER AND HOWLAND ISLANDS

With a sense of sickening horror, Margo gripped Klaus' hand as the awful sounds of the growing calamity at the socket reached them from across the water.

"Give me a hand, Klaus," Alex shouted as he secured Watson's hands and feet with duct tape. Klaus turned to assist while Margo continued to gaze back at Baker, now plunged into darkness.

The sickening feeling that swept over Margo was beyond description. Although she couldn't know the extent of the damage, clearly, the skytower was gone. *What about the skyport and Shift Three—Ohmygod! What a terrible way to go! And the loop....* She forced her mind to pull back. She tried to imagine the chaos at the socket. *The launch crew...Bruce!* Tears streamed down her face, unnoticed. "Oh, Alex...Klaus...It's so awful. All those people...the skyport...the loop..." To distract herself, she climbed into the Zodiac, loosened the line Alex had tossed across its transom, and tied it to the boat's tow ring. Klaus reached down to her and pulled her back onboard.

Skimmer One had aligned itself with the rollers moving from the north, illuminated by the full moon. The three stood in the stern well in stunned silence, gazing to starboard but seeing nothing. The moon reflected off the top of each roller as it passed, lifting them to where Margo could just make out Baker's darkened outline, and the silhouette of something ominously big protruding up from the island. Suddenly, without warning, the air filled with a whistling whoosh, and an unmistakable loud splashing sound to the south. Before Margo had time to make sense of it, the sound became deafening, and *Skimmer One* burst apart, sliced neatly in half just behind the pilot's stand. One moment Margo was standing in the stern well with Alex and Klaus in shocked disbelief, and the next, she was floundering in the water, watching the front half of *Skimmer One* slide down the backside of a moonlit wave and disappear.

Something nudged her. Margo twisted in the water to see the Zodiac several meters distant, still tied to the sinking back half of *Skimmer One*. Something nudged her again as a shiny dolphin snout emerged from the water by her face—it was George. "Oh George, George, you're here!"

A creaking groan drew Margo's attention back to the sinking skimmer threatening to pull the Zodiac down with it. Four strong strokes brought her alongside the Zodiac. She hoisted herself over the inflated gunwale with a friendly boost against her feet from George. With the full moon to guide her fingers, she undid the bowline she had tied through the nose ring just a few minutes earlier, and watched the after half of *Skimmer One* slip beneath the waves to follow the forward half to a watery grave. For a moment, Margo's thoughts turned to Amelia Earhart's final resting place, not that distant from *Skimmer One*. *You have company now, my Dear, something from me to share your silent vigil.*

"Margo…Margo…Alex…" Klaus' strong voice boomed across the water. "Say something, you guys."

"I'm here, Klaus, in the Zodiac."

"I'll be damned," Klaus shouted, "George is here with his buddies. Hey George…"

"Margo…Klaus…" Alex's voice slid along the water's surface, muffled by an intervening roller. "I've got one of George's buddies with me…can you hear me?"

"Coming, Alex." Margo cupped her hands around her mouth and directed her voice toward the water surface to give it more range. "Let the dolphin guide you."

Margo started the Zodiac engine with one easy pull and pointed the rubber craft toward Klaus' moonlit pate. A minute later, Klaus easily hoisted himself over the gunwale. He knelt in the bow and cupped his ears, listening for Alex's voice. "That way," he said to Margo, pointing off the port bow. "Okay, now straight…easy does it." Then Margo saw Alex and brought him alongside.

"Alex," Klaus said, but Alex did not answer. A dolphin's snout pierced the surface with a loud chatter.

"What is it, George?" Margo asked. George slipped beneath the surface and began to push against Alex's posterior.

"He's hurt. Give me a hand, Margo."

Klaus checked both Alex's arms, and then gently crossed them and pulled them across Alex's body, rotating Alex away from the boat. Margo took his right arm, and together, with infinite care, they eased him over the side on his back and into the boat. Margo checked his pulse and breathing. "Pulse and breathing okay," Margo said as George raised himself halfway out of the water and backed away from the Zodiac, chattering as he went.

On the deck, Alex groaned. Margo checked his legs for injuries and then gently moved her fingers across his shoulders and down his chest. As she reached the bottom of his ribcage, Alex groaned again. "I think he has a broken rib," Margo said.

"Oh shit!" Klaus said suddenly. "Where's Lars?"

BAKER ISLAND—BAKER COMPOUND

At the launch facility, Tom Daillebaust's first indication that something was wrong was when the lights went out accompanied by the ear shattering noise of the skytower self-destructing. The second indication was when a section of wayward cable crashed through the cafeteria roof, narrowly missing Daillebaust himself. The crash woke the sleeping Shift One crew. Mike Swamp, Pete LaFleur and the others stumbled bleary-eyed into the wrecked cafeteria.

"What the fuck happened?" Swamp asked Daillebaust.

"I don't know, but something's falling—we'd better get the fuck outa here!" Daillebaust shouted directions, and everyone in the cafeteria scattered through the southern door, running for their lives toward the beach. Illuminated by the full moon, plain for everyone to see, the falling skytower piled into an ever-growing writhing mass.

Daillebaust watched with a sinking heart as the double deflector smashed into a pile of cable that was already several tens of meters higher than anything within five thousand square kilometers. The gyro exploded into millions of pieces that peppered the entire island, including several people on the beach. Other than a few cuts and a couple of bruises, however, nobody was seriously hurt by the flying pieces. Daillebaust climbed the berm with Swamp and gazed around the devastated island—a shattered moonlit fairyland. The skytower cable whooshing northward caught his eye for a moment. He stood on the berm, stunned, shocked to his core, remembering a story from way back in 1907, when thirty-three Kahnawake Mohawks died during the collapse of the Quebec bridge over the St. Lawrence River. How many of my brothers died in this calamity? he thought, as unabashed tears flowed down his cheeks. Daillebaust raised his arms to the sky in a silent salute to the fallen. As he did, the bitter end of the skytower whipped through the air over his head, disappearing to the north, and silence settled over the island.

Daillebaust dropped his arms and looked at his friend, Swamp, with anguish in his eyes, and called out to the survivors on the sand below him, "C'mon, guys, let's see if we can find anyone else alive."

BAKER ISLAND—EASTERN BEACH

Bruce and Apryl really hit it off during their shared trip to *Amelia Earhart Skyport*. Although Apryl was the "older woman" in the relationship, Bruce was so very different from any guy she had ever known, that he fascinated her. Earlier this evening, they met on the eastern beach. Apryl brought some food and wine, and Bruce supplied the blanket. As the evening wore on, they finished the food and wine, and Apryl initiated Bruce into her sensuous personal world. The gentle onshore breeze kept them comfortable, and Apryl reveled in the certain knowledge that this night had changed Bruce forever. She had, she believed, spoiled him for other women, and that was just fine with her. She sat on his chest, stroking his face as they settled down from a round of passionate lovemaking. Tiny beads of perspiration glistened in the moonlight as they rolled down her breasts, dripping from her protruding nipples onto his nearly hairless chest.

Suddenly, the sky overhead flashed brightly. Apryl screamed. Bruce threw her to the blanket and covered her body with his. She felt Bruce push her against the berm, keeping his body protectively over her. Although she knew that she was much more experienced than he, somehow his actions touched her, and she felt protected, despite herself. As they huddled against the berm, their world crashed down around them. Several lengths of cable snaked over their heads, coming to rest partly in the water.

When the double deflector crashed into the writhing pile, Apryl screamed again, and tucked herself even tighter into Bruce's arms. "It's okay, Baby," he told her with husky tenderness. "It's okay."

But it wasn't, and she knew it wasn't. She contemplated the irony of tough, capable Apryl Searson—afraid of no one, torrid lover who chose partners like a man, but loved like a courtesan—taking succor in the arms of a boy. They stayed huddled together, terrified and alone for nearly an hour.

Apryl roused at the sound of voices from the other side of the berm. She thought she recognized them. "Mike, is that you? Tom…?"

"Apryl?" Swamp called out through the moonlight.

"It's me, Mike…Tom."

"What the hell, Apryl," Swamp said as he topped the berm. "Oh shit, you're naked!" Bruce stuck his head out of the cover. "Bruce! You and Apryl? Shit, man, I'm sorry…c'mon, get dressed. We gotta get you guys off the beach, and we gotta find the other survivors." Tears streamed down his face. "Oh shit…Oh shit!"

BAKER ISLAND—BAKER COMPOUND

Lori was going over her day's notes in her room with Mary Martain from *Me TOO! Magazine*. Because of her heroic actions on *Aku Aku* that so materially assisted in saving Bruce's life, Alex had granted Martain permission to come to Baker for some personal interviews. When she arrived a day earlier, Lori immediately took the cute, young writer with bobbed brunette hair under her wing. By the end of the first day, Lori found herself fondly recalling her night with Margo, and wondering if she would repeat those sensuous hours with this exciting young woman.

Following dinner with the Shift One crew in the cafeteria, Lori invited Martain to her room to review the day's notes. Two glasses of wine and the warm island air had their effect. In the tropical closeness, the women shed their outer clothing and then their inhibitions. Lori led the younger woman to the bed where they collapsed, body to body, arms and legs intertwined. Once again, Lori marveled at the sweet softness of a woman's lips when compared to the men she had loved. She parted her lips…Martain responded…and their entire universe exploded around them. One moment they were lovers in the dark, the next, Lori was thrown to the floor, the young woman torn from her arms.

Lori groped in the dark on her hands and knees, searching, calling out. "Mary, where are you? Mary…answer me…Mary…" Then her hand slipped on something wet and warm. She recoiled, and then probed further. She grasped an ankle in the dark and followed it up the firm young body she had just made love to. She placed her ear next to the lips she had kissed only moments before—no breath. She laid her face against a breast she had so recently caressed—no heartbeat. "No…" she moaned softly. "Not this way…" She cradled the young woman's still form against her breasts, sobbing quietly.

✺

Apryl Searson found Lori an hour later, sitting unclothed on the floor, holding the nude body of the younger reporter against her own, rocking back and forth, and sobbing quietly. Apryl placed her lantern on the floor and checked Martain's pulse, finding what she knew she would find. "Lori," she said softly. "Lori, listen to me." Apryl gently removed the still-warm body from Lori's arms. "She's gone, Lori." She draped a shirt across the reporter's shaking shoulders. "We need to get you dressed, girl." She boosted Lori to her feet and sat her on a chair, since the bed was smashed. "Here, slip this on." She lifted Lori's feet one at a time and slipped one of the thongs lying on the floor over her hips. "Looks like you girls had a fine time," she said solemnly. Apryl buttoned the shirt. "Don't worry about the bra," she told Lori. "You'll do fine without it." She pulled the stunned reporter to her feet and clasped a short skirt around her waist. Then she pushed sneakers on the woman's feet.

When Lori was back in the chair, presentable to the outside world, Apryl turned her attention to the beautiful young reporter's lifeless body. Her probing fingers discovered the back of the young woman's skull battered in. *Looks like you took it for Lori, kid. If you were not on top of her, I'd be probing Lori right now. Too bad…you were one hot chick.* With professional detachment, she slipped shorts over the girl's hips and a t-shirt over her torso. She ditched the bras and extra thong, and laid a sheet over the beautiful young reporter's still body.

"Now we'll call the others," she said, putting her arms around the sobbing woman. "We'll keep this between us." She kissed away a tear from Lori's cheek, and wiped one from her own.

Dex burst through the door, holocam lights flooding the room. "Lori…are you okay?"

"Not really," Apryl told him. "Mary's dead."

"Oh…." Dex's face fell in dismay. "I liked her a lot."

"So did Lori," Apryl said, holding Lori closer as she sobbed her heart out.

CHAPTER TWENTY-SIX

SEATTLE—SMITH TOWER

Mabel Fitzwinters sat at her ornate desk atop the Smith Tower in Seattle, holodisplay floating before her, perusing the Web as was her wont each morning. It was still early. She was the only one in the *LLI* offices. Yesterday had been particularly full, since she had stayed on top of the first launch pouch and kick thruster shipment as it crossed half a world to the socket at *Baker Skytower*. She was on the threshold of fulfilling her dream of creating an inexpensive doorway into space and the universe beyond the confines of Earth's atmosphere. She could no more sleep than she could stop breathing. She was as excited as the day so long ago when Tommy popped her cherry, as the day she accepted her doctorate at Stanford, or the day Launch Loop International became a reality.

She sipped a coffee produced by the copper-clad antique coffee maker that occupied a prominent corner of her suite. *Seattle might be the coffee capital of the world*, she thought, *but nobody makes better coffee than this wonderful old machine.* She leaned back, savoring the coffee's rich flavor, when her holodisplay flashed red, and Alex appeared in a corner of the display. She signaled her Link, and

Alex displaced her Web view, as it shrank to occupy the corner Alex had vacated.

Alex was disheveled, and his clothing appeared soaking wet. "We got a problem, Mabel. A big one."

※

Quinton Radler sat at his functional desk in the fourteenth-floor corner office of the Smith Tower. *The Environment, Inc.*, CEO stared intently at a holodisplay hovering over his desk that mirrored Mabel's display twenty-one floors above him. In a small section of his display, a cyberbot disgorged the running results of its search as it scoured the Web for any breaking news about Slingshot or *LLI*.

Radler sipped on a latte he had picked up at the boutique coffee stand in the lobby when he arrived a few minutes ago. The Fitzwinters broad appeared to be doing what she usually did each morning, and Radler paid it only slight attention, concentrating, instead, on the cyberbot output. When his display flashed red, however, he gave it his full attention. Alex Regent had just said, "We got a problem, Mabel. A big one."

Radler listened with grim satisfaction as Alex briefed Mabel. "The second capsule carrying kick thrusters exploded at the terminus. The only reasonable explanation is a bomb, and we think we know how it happened. The skytower collapsed, killing the launch crew—all three of them. We haven't found their bodies yet, but they had zero chance. The launch facility was completely obliterated—there's nothing left. Do you remember Mary Martain, the reporter from *Me TOO!* who helped Bruce escape? She was on Baker doing some interviews." Mabel said she remembered. "She was killed. We have some injuries in Shifts One and Two and the support personnel—nothing serious, though. *Earhart Skyport* appears to be undamaged, and Shift Three is stranded there. They're mostly unhurt, except for Rudolph Pigman. He broke his arm. Pigman was a replacement for Kelly Seidell, the guy with the meteorite puncture. We're working on a way to bring them down. Margo, Klaus, and I were on *Skimmer One* chasing down, as it so happened, Lars Watson, who had infiltrated Shift One. We believe he placed the bomb and then hightailed it for Howland in a stolen Zodiac. It's complicated, but I'll brief you on the details later. Right after we captured Watson, the falling cable sliced *Skimmer One* in half. We're okay, though." Alex winced.

"What are you not telling me, Alex?" Mabel asked with concern.

"Nothing really. I broke a couple of ribs, is all. Apryl Searson, one of our diver medics, taped me up. I'll be fine, Mabel. I just need to avoid exerting myself for a few days."

"And Watson?"

"We had him tied up, and when *Skimmer One* sank, we lost him. Margo's dolphins kept him on the surface and led us to him. We have him in detention. Except for some cuts and bruises, he's unharmed."

"So…obviously the loop survived," Mabel said.

"Yep…No damage at all up there that Critz can detect. Those gyros you built really did the trick." Alex grinned at her, and then winced again. "Gotta remember not to do that," he muttered. "Soon as we figure out how to get those boys down, and as soon as we come up with a way to reconnect the skytower, we'll be back in business." Alex turned away from his holodisplay, apparently listening to someone off-camera. "Mary Martain was with Lori when she was killed. They had what I guess you might call a special relationship. She's pretty shook up, but she insists on making a live broadcast to the world as soon as I release her." Alex smiled wryly and winced again. "I think it's a good thing. It's time the world discovers the true nature of *Green Force*…" Alex turned away from his holodisplay again. "Something's come up, Mabel. I'll get back to you shortly."

"Hold it, Alex—just a moment." Mabel did something off-screen. "I'll be at King County Airport in forty minutes," she said, "and on Howland as soon as I can get there."

Radler's display reverted to his normal holding pattern, and he established a link to Darius Gotch. Radler briefed Gotch on what he had just intercepted, finishing with, "Can't your guy do anything right? The loop is still running, and you know they'll find a way to put up a replacement skytower." Radler scowled at Gotch. "You won't have another chance with a bomb. In fact, you won't be able to get anyone near the project anymore."

"We need to rethink this," Gotch said.

"Yah, right…take a week. Muster your resources, and come back to me with another plan. One that'll work."

PYONGYANG– DPRK (NORTH KOREA)

General Jon Yong-nam leaned back in his lushly padded office chair deep in the sub-basement bowels of a squat government building overlooking Kim Il-sung Square, a block west of the Taedong River. He was staring at a holodisplay over his ornate desk, replaying Radler's hack of Mabel's system, reviewing Alex's report. Unlike Radler, The General was pleased with the outcome. The fall of *Baker Skytower* would suck up huge LLI resources, and set their timeline back significantly. Let Radler stew in his failure while his people devised a better method of bringing down the rail itself.

The general fiddled with his link controls, and in a moment, the walls of his office disappeared behind a holographic display that mirrored the walls and windows in Mabel's quarters atop the Smith Tower in Seattle. He nodded to himself as he gazed through the virtual windows. *That's a fine view of a beautiful city.*

As the image of Alex collapsed into itself, the General called up the latest launch booster order summary. Nearly one launch every two days for the next two months. He smiled to himself. At ten million U.S. dollars per launch, that was nearly $300 million badly needed cash for his beloved DPRK—and a tidy ten percent for his own pocket. *And that old fool, Kim Jong-un, has no clue.*

The general arranged for another substantial deposit into an offshore account that would eventually end up in the coffers of *EI*. Chump change compared to the profits he was reaping from his booster activities. He hesitated, and then performed another transaction, swelling his personal account by several million.

Before retiring for the day, the general adjusted his controls again. This time the holographic image that replaced his walls mirrored the ornate interior of a room high atop the government palace in central Teheran, the private office of the Persian Caliph. For a few minutes, he let his eyes roam over the ornate fixtures, the priceless carpet, the ancient wall hangings. "You and I will do business soon," he whispered quietly as he shut down the display and departed for the day.

SEATTLE—AIRBORNE TO BOEING FIELD

Mabel climbed into the waiting helicopter on the twenty-first floor observation deck, and immediately established a link to Alex. "I'm in the chopper on my way to Boeing. Let's finish our conversation."

"I cut you off before, Mabel, because Carey Endsley had just hacked into your system…" Mabel started to interrupt, but Alex held up his hand. "Wait a sec, Mabel. Carey hacked into your system, and found someone there before him. Carey traced it to Quinton Radler, CEO of *EI*, down on the fourteenth floor. Carey says they're hard tapped into your system."

"My God," Mabel said, astonished and troubled beyond words.

"That's not all," Alex continued. "First though, you need to contact all your senior people, and tell them not to use the company Link for anything sensitive. Tell them to use only encrypted personal Links for sensitive stuff. But," Alex emphasized the next few words, "make sure they continue to use the company Link for routine matters."

During Mabel's short flight to King County Airport, and then during her extended flight to Howland, Alex brought her up to date on his preliminary findings and where they seemed to be taking him. He told her that he was fairly certain *EI* was receiving its funding from *a government entity in North Korea*, but he didn't have the resources to follow it any further. "Knowing it, and proving it in court, however, are two different things," he told her. "We're working on that. Carey can do things on computers and in the Web that I didn't think were possible. We follow the money, and I think we can get these guys on murder charges. You better bring John Boyles into the picture. We'll get him the evidence, and if anyone can pull this off, he can. One more thing," he told her. "It's critically important that we don't let *EI* know what we've learned."

Mabel spent the rest of her trip planning their attack by Link with John Boyles and Rex Johnson. They suggested that Carleton Montague and Delmer Woodward be brought into the picture. "If we're going to beat these guys," Boyles told her, "we're going to need all the help we can get."

BAKER ISLAND

Sunrise brought with it the full extent of the devastation on Baker Island. The pile of cable on top of the socket consisted of sixteen kilometers of tangled boost cable, less several hundred meters that were shredded as the cable shed its momentum from its 400-meter-per-second velocity. This shredding process also destroyed the immediate launch facility, killing the three men on duty. Their bodies never were found, although for days after the crews untangled and removed the cable, sea birds flocked over and about the socket, finding morsels of human flesh that they carried off to their mates and nests. Alex found it difficult to look at the birds the same way after that, even though he knew, intellectually, that they were simply doing what sea birds do.

Because of his broken ribs, Alex was forced to curtail his supervising activities. Klaus assumed the task while directing Apryl to put Alex to bed and keep him there for a couple of days. Alex protested, but not too much, knowing that Klaus would do just fine. He reached a compromise with Apryl, and set up a workspace that he could use while sitting, propped up in bed. Klaus rigged several strategic holocams so that Alex could follow most of what was happening outside.

The first task was removing sixteen kilometers of cable. Although the cable appeared undamaged, Alex was reluctant to reuse it, given the circumstances under which it had reached the ground. After considering the alternatives, Alex and Klaus decided to use the ocean-going tug to tow one end of the cable to sea, going south, until the bitter end was several kilometers south of the island. The solution was simple, elegant, and had virtually no ecological impact. By 8:30 AM, teams had dragged the cable end to the water's edge, and attached it to a line from the tug. Five minutes later, it was on its way out to sea. The tug maintained a ten-knot speed, and a scant hour-and-a-half later, the cable was just another minor bump on the abyssal floor, a bump that would disappear under a layer of silt within a year.

※

Early that afternoon, after a thirty-nine-hundred-nautical-mile supersonic trip, the *LLI* corporate jet touched down at *Amelia Mary Earhart International Airport.* Mabel Fitzwinters transferred immediately to the waiting Chinook, and twenty minutes later, she disembarked on the Baker tarmac following a short circumnavigation of the island. Margo and Klaus met Mabel as she stepped onto a frying-pan-hot surface under a blazing mid-afternoon sun. The sprinkling of white clouds did nothing to mitigate the sun's intensity. As the Chinook twin blades coasted to a stop, several dozen birds wheeled into a disorganized cacophony of sound and sight, soaring and plunging as they rode the currents rising from the tarmac. Mabel looked around in amazement. She had expected to see the enormous pile of cable rising from the socket, but on her flight around the island, all she saw were birds fussing over a few destroyed buildings. *Despite the tragedy*, she thought, *this is a tropical paradise, a bit hot perhaps, but a paradise nonetheless.*

"Where's Alex?" she asked.

"In bed—Apryl's orders," Klaus said with a twinkle in his eyes. "She's being a real mother hen about him."

"I want to meet this girl."

"We're headed there now," Margo told her as they climbed into the waiting pickup.

In Alex's room shortly thereafter, Mabel made herself comfortable on the couch someone had hastily dragged into the room. Cool, conditioned air eased the discomfort she felt from the unaccustomed tropical heat. She gratefully accepted a rum-spiked cooler that Noah, the Baker Compound Steward, pressed into her hand. Bell-like voices of Maori girls floated through the air, just above the threshold of hearing. She looked at Alex warmly. "Thank God you're okay." She turned to Apryl, who stood shyly to one side. "You, child…" Mabel motioned her closer.

"Yes'm."

"None of that. You're doing a wonderful job…," Mabel ran a quick mental name-recall routine, "…Apryl. First Kelly Seidell, and now the boss himself. It appears you have made a bigger difference than we ever imagined when Emmett Bihm hired you." Apryl blushed and stepped over to Alex to fluff his pillows.

Noah slipped noiselessly into the room and whispered into Klaus' ear. Klaus turned to Mabel. "Lori from the Fox Syndicate is requesting a short interview with you here in Alex's room. She's putting together a special on what happened, and I think she wants to include some footage of your visit with Alex."

"Send her in, of course," Mabel said, looking forward to meeting the woman who, despite her severe emotional shock just a few hours ago, was on the job, doing her professional best.

Lori entered the room with Dex Lao in tow. She wore a restrained gray skirt and white blouse, and had swept her hair onto her head to keep it out of her way during this hectic day. Margo stepped over to Lori and squeezed her hand, mouthing something that Mabel could not make out.

"Mabel Fitzwinters, this is Lori Kutcher from the Fox Syndicate." Margo led Lori to the couch. "Lori, this is the woman who made all this possible."

"I hope you don't mind if I remain on the couch, Ms Kutcher," Mabel said. "I'm feeling the tropical heat."

"Lori, please…," Lori said as she leaned over to kiss Mabel's cheek.

What an utterly charming young woman. Mabel smiled warmly at her. "I'm so sorry for your personal loss." Lori blushed slightly.

"We will all miss Mary," Lori said. "She became pretty important to all of us when she helped Bruce escape from the *Aku Aku*, you know."

Hardly noticed at the side of the room, Dex recorded the conversation.

"Mabel Fitzwinters," Lori said more formally, "would you please tell my viewers why the Chairman of Launch Loop International came all the way down here on such short notice."

"This is an unusual business, Lori. We've just suffered a cataclysmic tragedy. These people," Mabel swept the room with her arm, "are like my family. They don't need me here, of course." She chuckled wryly. "In fact, I'm probably in the way and consuming valuable resources and personnel time…" She held up her hand as Margo started to protest. "But there are times when face-to-face is the only way. I can't give a needed hug over a Link. I can't wipe away a tear long distance." Mabel touched the corner of her eye with a napkin. "Right here, right now, I'm needed. I won't overstay my welcome, but I have to be here when I'm needed."

※

A half-hour later, Lori smoothed her skirt, squared her shoulders, and set her features to their most professional look. "We are standing at the Baker Socket where, just a few short hours ago, the *Baker Skytower* rose eighty kilometers into the sky to meet with *Amelia Earhart Skyport*. This is the western end of a twenty-five-hundred-kilometer-long doorway into space that will begin commercial operation in a few months—although that start date has had to be pushed back as a result of a major setback the project experienced last night. Even as I speak, seven brave men are stranded eighty kilometers above us in the sky over Baker Island. Alex Regent and his people are frantically working to rescue them." Lori described what had happened the night before, and then reviewed the events that led up to the explosion. She was specific and graphic when describing the role *Green Force* played. She included an interview with Bruce where he described his infiltration of *Green Force*, and what he learned about their goals and tactics. She included another interview with Margo, where she described how *Green Force* had cut the tensioner cables, and her efforts to put them back in place. She concluded with an analysis by Alex from his bed of what might have motivated *Green Force* to take such drastic actions, of what it must have cost to put the entire plan into effect, and why all the information pointed to a behind-the-scenes party with deep pockets and a lot to lose from Slingshot's success. During his analysis, Alex carefully avoided mentioning specific names, but his implications were unmistakable. *The Democratic Peoples' Republic of Korea's* fingerprint was everywhere.

 Lori finished up. "The huge pile of cable is already gone. The wreckage of the destroyed buildings has been removed. Rescue efforts are underway to bring the seven stranded men back to Earth. *LLI* is gearing up to make a replacement cable that will be ready in a few days to load onto barges like these at the Port of Seattle." The scene shifted to stock footage of several ocean-going barges. "By the time the cables arrive a week later, the buildings will have been replaced, and the launch facility will be ready to receive the cables. With a bit of luck and a whole lot of sweat and hard work, Slingshot will be ready to move forward a month from now, God willing." Lori smiled her best professional smile. "This is Lori Kutcher with Fox Syndicate coming to you from Slingshot on Baker Island in the equatorial Pacific—the new gateway to the stars."

CHAPTER TWENTY-SEVEN

SEATTLE—SMITH TOWER

"I can set us up for another shot," Darius Gotch told his boss, Quinton Radler. "We have an open window with a tight timeline." They were sitting at the conference table in the Smith Tower fourteenth-floor offices of *Environment, Inc.*, with personnel head Dane Curvin and Katelynn Leete, who implemented manufacturing sabotage.

"I'm listening," Radler said.

"They have to check every inch of *Earhart Skyport* before they certify it for passengers. If we can force them to accept a government inspector as part of the team, perhaps we can insert Watson one more time."

"Why Watson? He's screwed up twice already."

"Not really, Quinton." He held up a hand as Radler appeared ready to interrupt. "I know the job didn't get done, but Watson DID cut two tension cables. We learned a great deal about their capabilities from that. The bomb did bring down the skytower. But it wasn't Watson's fault that the bomb wasn't strong enough to crash the whole thing. He did his job. It was pure bad luck that he got caught."

"Don't need that kind of luck," Radler said.

"I know, but hear me out. If you can get him on the inspection team as a government weenie, here's what he can do."

Gotch opened his case and pulled out a cylindrical object that he placed on the conference table. "This is a section of the sheath that holds the rail. It's the identical material. Watson—as the government weenie—will be required to inspect the sheath closely anywhere near the point of explosion." He reached into his case again and pulled out two glass vials of colorless, water-like liquid. "This is some amazing shit. Watch." He snapped off the tops of the vials and emptied them onto the sheath section, where they formed a viscous gel. The area covered by the gel appeared to do nothing at first. Gotch started a timer and then watched while they talked. Five minutes passed—still nothing. Another five minutes passed while the people at the table continued to discuss the overall problem. Then, as if on cue, at the ten-minute mark, the surface shriveled rapidly and fell into the sheath. "There's something going on you don't see," he said. "This stuff works like acid on the sheath material. It is designed to subsume the surface layer over ten minutes, and then it eats right through the rest of the sheath. When it reaches the iron rail segments, however, it's a whole different ball of wax." He extracted a piece of rail from his case—three interlocking segments, the middle one a double male. He demonstrated how the segments slid together and apart, almost without friction. He sprinkled the residue from the disintegrated sheath onto the male section right where it slid into a female section, and then moved the segments together. When he tried to pull them apart, he could not. "Try it," he said, passing the segments around the table.

"The residue," he said with a grin, "creates a molecular bond between the soft iron pieces, so they fuse together. Put enough of this shit on the sheath at Baker, so you fuse enough sections, and when they reach the Jarvis deflector two-and-a-quarter minutes later, they tear the rail apart. Since they can't bend, they try to continue straight on—and rip the deflector to pieces, and with it, the rail." Gotch leaned back with a self-satisfied expression on his face. "And look at this." He took out two more glass vials, opened his bottle of drinking water, and dropped them into the bottle. They disappeared. "Both liquids have the same refractive index as water. Our guy can take as much as he needs with him. No one will ever know."

BAKER ISLAND—BAKER COMPOUND

"My old college buddy, Judge Kalolo Leaupepe of the American Samoa Federal Court, called me this morning," John Boyles told Mabel Fitzwinters on her personal Link.

"American Samoa Federal Court? I didn't know they had one," Mabel told him.

"For years now," he said. "Part of the ninth circuit. Anyway, his call was a confidential heads-up to an old friend. *Green Force* is about to file a *writ* to spring Watson. He says they'll succeed, although we might be able to hold them off for a couple of weeks. He suggests that we officially request the Seattle Federal Court for federal marshals to escort Watson back to Seattle. Says we need to do it immediately, today, even."

"Okay, John, see to it. I'll inform Alex." Mabel closed the Link.

What's your game plan, Mr. Radler? What devious scheme are you cooking up now? Mabel called Alex, telling him about the marshals. "They should be there in a day or so. Release Watson to them, and good riddance."

❇

Klaus and Margo sat on the couch in Alex's room since Apryl still insisted that he remain in bed. "So…what are our options?" Alex asked.

"Three, as I see it," Klaus said. "They jump, but it's never been done from so high, and I don't think they have the right equipment anyway. That's on my list for things to do tomorrow." He grinned. "We lower the rail until we can remove them with the Chinook—ours has a ceiling of about nine thousand meters. We've got to bring it down to attach the new skytower, but we're not ready yet, and Critz and his boys are getting thirsty and hungry." Klaus spread his hands in the air in front of him, palms down, fingers spread. "Or, we send them a capsule from Jarvis, and they ride it back."

"We haven't tested any of that yet," Alex said.

"So we run our tests on the spot," Klaus answered. "We run a capsule from Jarvis and send it back. We do it several times to iron out any kinks, or to find them and then iron them out." He sat back and smiled. "It'll actually speed up the overall process, since we will have to run those tests anyway."

"He's right, Alex," Margo chimed in. "The tests are entirely independent of the skytower. The only thing we can't do is send the capsules down to Baker once they arrive at *Earhart*."

There were two methods for transferring a capsule between skyports. System design called for launching a capsule from *Noonan* down the rail with the kick thruster, while using the rail to brake it at *Earhart*, and carrying a capsule backward down the launch rail from *Earhart*, and braking it with the kick thruster at *Noonan*. The *Earhart* to *Noonan* transfer carried the greater risk, since a failure of the thruster could cause the capsule to crash into the receiving skyport if it could not be jettisoned from the rail in time. They continued the discussion, tossing numbers back and forth, exchanging holodisplays, arguing specific procedures, but in the end, even ignoring the fact that the reverse thruster was the only way to get Shift Three back to *Earhart*, it looked like a combination of both procedures would work.

"Critz, it's Alex. How're you guys doing up there?"

"Considering, not too bad. We're getting a bit tired of candy bars, but the water is holding out, and so are we. You guys come up with something yet?"

"Actually, that's why I called." Alex outlined their plan to Chrietzberg. "Give us another day or so, and we'll be ready to send you a capsule." Alex gave him a wry grin. "We'll try to have some real food in the first capsule."

"We're running out of pain relievers for Pigman, Boss."

"We'll send some as well." Alex closed the Link.

"Okay, guys, road trip time." He hoisted himself out of bed, wincing a bit. Apryl rushed to his side.

"You're not ready yet, Boss. Seriously, you need a couple more days."

"Can't do it, Apryl," Alex said, winking at her. "Love to stay with you, but Critz and his guys need me." Alex slipped on a shirt, stifling another wince.

"Take it easy, Boss," Apryl told him. She turned to Margo. "Please keep an eye on him, Margo. His ribs haven't healed yet."

"Klaus, would you please have the guys ready *Floater One*? We can get underway in an hour."

"Ain't gonna happen, Boss," Klaus said with gruff tenderness. "You're in no condition to pilot a plane two thousand klicks. I'll have

Tex bring *Floater Two* out here, and we'll leave in the morning. Critz and his boys can hold out a bit longer." The big German took Alex's shoulders, turned him around, and gently picked him up and laid him back on the bed. "Now you stay there till morning, my friend."

Apryl beamed and raised herself up tall to kiss Klaus on his cheek.

※

Early the next morning, a chartered business jet landed at *Amelia Mary Earhart International*. Their flight plan was filed in Seattle, and the pilot identified himself as under a federal charter, carrying federal marshals. His passengers, he informed the control tower, required immediate transportation to Baker Island to pick up a priority prisoner. The tower called Alex, and he authorized the Chinook to bring the marshals to Baker. By 7:30, Alex was talking with the two marshals in his Baker office. The marshals identified themselves as Deputy Marshals Lance Banes and Allan Hutshell. Their credentials were in order, and they had the look and feel of federal agents. They presented Alex with transfer documents, and he scanned both the warrant and the agent's documents. Then Alex officially turned over Lars Watson to the federal government.

The marshals cuffed Watson and walked him outside, where they were driven to the tarmac. Thirty-five minutes after they left Alex's Baker office, the federal charter rolled down the Howland runway and surged into the tropical sky.

JARVIS ISLAND—JARVIS COMPOUND

An hour later Tex Sanchez fired up *Floater Two*, carrying passengers Alex, Margo, Klaus, and Apryl—who had insisted on accompanying her patient. Their four-hour journey promised to be routine. Radar indicated no weather systems, not even any squalls along their route. Alex was sore. Even the relatively slight push from *Floater Two's* acceleration caused him to wince. Apryl picked it up immediately, and fussed over him, attempting to make him more comfortable, much to his amusement. He marveled again at his luck of the draw in putting together the Slingshot team. *Take Apryl*, he thought. *She was a young diver with ambition and a yen for adventure—like thousands of others. But she rose to the unanticipated challenges we threw at her.* He examined her out of the corner of his eye. *Pretty as a picture, too!* Alex looked out the window at the passing waves below. It felt odd not doing the flying. He let his mind drift over the problem ahead. He trusted the loop. Thus far, every element had worked flawlessly, and he had no reason to doubt that their initial test of the east-west travel feature would work fine. He was concerned that they were way early doing this. They were bypassing a whole series of planned intervening steps. A call from the Howland tower interrupted his reverie.

"Mr. Regent, we have an inbound from Seattle. Says they're federal marshals with orders to pick up a priority prisoner on Baker."

Alex placed an immediate encrypted call to Mabel's personal link. "Mabel, please get the names of the marshals the court sent to pick up Watson. I need those names in the next few minutes." Mabel called him back in about fifteen minutes with the names: James Franks and Ronald Botts. Alex called the tower back. "Have the marshals met at the gate with my Link coordinates. Ask them to call me."

Shortly thereafter, Alex received a call. "This is Deputy Marshal Jim Franks. I've got Deputy Marshal Ron Botts with me. We are carrying a federal warrant to pick up Lars Watson on aggravated terrorism and murder charges."

Alex explained the earlier pick-up to them, and transmitted a copy of the IDs and warrant the impostors had used.

"Those IDs and warrant are genuine," Franks told him. "Wait a sec..." While he waited, Alex worked to control his rising rage. Franks continued, "Banes and Hutshell worked out of the Pago Pago office until two days ago. They resigned and fell off the radar."

"Any way to find them? To tell whom they work for?"

"It's a big world, Mr. Regent, and they obviously have money behind them. We'll do what we can. Sorry they got your man, sir."

As Alex sat in his seat, fuming over the snatch, the Howland tower called him again. "Sir, we just ran a check with Seattle. That first federal charter did not come from Seattle, sir. It came from Samoa."

❋

Alex's first order of business was Chrietzberg and his shift. As soon as *Floater Two* landed at Jarvis, however, Apryl insisted on putting Alex on a bed while she examined him and changed his dressings. Alex was about to override her when Margo caught his eye. He changed his mind. Apryl needed to know that her input mattered, and besides, she probably was correct. He lay back against the pillows and let Apryl minister to him. She gently removed the tape around his lower ribcage. The discoloration had diminished, but Alex still was sore. Apryl bathed his chest, gently dried him, and carefully wound new tape around him. She stepped back to admire her work, and Alex thanked her and said, "Gotta get moving. The capsule's waiting for us."

"Not us," Apryl said, "them. You're not going on the capsule. It could kill you!"

Alex started to protest when Margo stepped into the room. "She's right, Alex. And you know it. We got enough heroes around here." Margo smiled and gave him a gentle hug. "Speaking of heroes," she added, "somebody wants to see you." She waved at the door.

Bobby Pfaff stepped through the door and smiled at Alex shyly. He was cleaned up, without piercings, and he looked strong and healthy. "Bobby…what a delightful surprise!" Alex shook his hand warmly.

"I'm, like, clean, Sir, and it turns out I am good at something." Bobby presented him with a wide smile. "Tex taught me about, like, airplane engines, and it turns out, I'm good at that. I mean, like, I'm really good! I understand them—and not just airplane engines…"

"Whoa…slow down, Bobby! You can tell me all about it on our way to the socket." Alex laughed and put his arm around the boy's shoulders. "You know what's going on, Lad?"

"Sure do, Sir."

They walked out the door into the tropical heat of Jarvis Island at noontime. The humidity was up, making breathing more of a task than normal, adding to Alex's broken rib complications. The birds didn't seem to mind, though. They were all over the sky, as if they were welcoming Alex back.

"That's a sooty tern," Bobby said, with a laugh, pointing at a swooping bird. "And that's a masked boobie." He practically danced beside Alex in his excitement. "Did you know that one," he pointed to a lesser frigatebird, "gets a lot of its food by, like, robbing the other birds?"

"That so?" Alex could barely contain his amusement.

"Yep…I, like, see it every day. It's cool as shit. They dive-bomb their prey."

Sanchez hurried up behind them. "Alex, Bobby…so we're ready to do it?"

"Hey, Tex…" Bobby waved at Sanchez. "Tex has been helping me a lot," he said to Alex. "Remember, you asked me once whether I was ready to give up all that *Green Force* crap?"

"Sure, Bobby, and…?"

"Like, yes!" Bobby laughed out loud. "You guys, like, rock!"

Everyone with nothing else to do was waiting at the launch ramp when Alex and his troupe arrived. The capsule was loaded with seven hot meals in a thermo carrier. Alex decided to send only Margo and Klaus. Jarvis Shift One was already in the *Noonan Skyport*, so they would have all the help they might need.

The countdown proceeded without incident, and five minutes later, Klaus reported, "Okay, Alex, we're here…sorry you can't be here." He paused. "Enjoy the ministrations of that little medic. We should all be so lucky." Alex was certain he heard a chuckle at the end of the transmission.

CHAPTER TWENTY-EIGHT

FRED NOONAN SKYPORT

Klaus had not been up the *Jarvis Skytower*. It was near noon, so the sun's position gave no directional indication. His instinct told him that the capsule dock was to the south, but in this case, instinct failed him. It was as if he were looking at an Escher staircase, waiting for the perceptual shift to turn it upside down. Internally, *Fred Noonan Skyport* was virtually identical to *Amelia Earhart Skyport*. Since *Noonan* was primarily a receiving facility for capsules sent from *Earhart*, there were some external differences. *Noonan* had external docking facilities for holding several capsules serviced by a second gantry. The docking facility included an automatic method of stripping an arriving capsule of its spent kick thruster and launch pouch. The logistics for returning the retrieved accessories to service was not yet worked out, but it was a big deal that still needed to be addressed in detail. At full capacity, Slingshot was capable of retrieving an incoming capsule every ten minutes, so the receiving dock required a significant amount of thought.

Klaus stepped onto the main deck, soaking in the environment, mapping it in his brain so that in an emergency, he could react to that pattern even if vision were impossible. More than once in his life, this practice had saved his butt. Shift boss Ajay Dybo walked up to Klaus. Dybo was Tom Daillebaust's cousin—they shared the same Mohawk grandfather, and worked on high-iron projects together the world over. Fellow Mohawk Prajeet Kahnawak and Chicago tough guy Erik Maffet were right behind him. "Welcome aboard, Boss," Dybo said.

"AJ…" Klaus shook his hand and those of the other Shift One guys. "Give us a few minutes to work out the details of our plan, AJ. Regardless of our final approach, we'll need our capsule outfitted with launch pouch and kick thruster. Let your automatics do it, but give it a personal check since this is the first time."

"You heard the man, guys!" Dybo said. He grabbed Kahnawak and left to suit up. "The rest of you know what to do."

"The way I see it," Margo said to Klaus as they sat at a table by one of the windows, "is that we are really using the rail as a continuous brake against the kick thruster."

"That's what this entire operation is all about, Margo—precise insertion into a specific orbit, and precise control on the rail." He grinned at her, knowing he was not telling her anything she did not already know. "From the controlling computer's perspective, one complex operation is no different from another. If any element fails, the whole thing fails."

"How long will the rail and thruster to *Earhart* take?"

Klaus ran some numbers on his Link. "Assuming a one-gee acceleration to the half-way mark, and a one-gee deceleration for the balance…about a half-hour. Double the gees…about twenty minutes. At three gees…sixteen minutes."

"That's too much. What about one and a half gees?"

"A bit over twenty-two minutes." They looked at each other, and Klaus smiled. "I think you made your point. Let's set it up for one point five gees. Do we agree?"

Klaus briefed Alex on their conversation. Alex concurred with their analysis. Klaus set the capsule parameters so it would float free on magnetic suspension, accelerate with the kick thruster while simultaneously being braked by the rail to produce 1.5 gees for 660 seconds, and then decelerate at 1.5 gees using magnetic friction guided by radar to come to rest at *Amelia Earhart Skyport*. He called Chrietzberg. "Critz, how are you guys holding out?"

"Okay, I guess. We just finished our last candy bar. Water supply is running out; the holding tanks are near capacity."

Klaus briefed him on their decision. "I want you guys suited up when the capsule arrives…just in case," Klaus said. "As soon as you unload your dinners, send the capsule back. We want to give the return trip a dry run before committing you guys to it."

"We're not going anywhere in the meantime."

Klaus turned to Dybo, who had just stepped through the lock onto the main deck, still suited up, with his helmet under his arm. "You guys ready?"

"Checked and double-checked, Sir…" His inflection left a question hanging in the air.

"What is it, AJ?"

"We've never seen one of these kick boosters burn. How intense is the flame? How long will we feel it?" The other shift members nodded their heads. Klaus could see that this was a major concern for these tough high-iron men.

Klaus chuckled. "That's the crucial question, isn't it, at least from where we sit?" Everyone nodded. "You all have participated in the construction of this skyport from the beginning, right?" Again, they all nodded. "Then you know the big difference between the launch bay and the capture bay." The shift members looked at each other. Klaus continued. "Since the launch bay uses only magnetic coupling, the skyport end of the bay holds the attaching mechanism for the kick thrusters. All the capture bay has against the skyport is a smooth, shaped surface…right?" They nodded again. "That surface directs the thruster blast away from the skyport. For the short time the thruster blast strikes it directly, it can withstand the heat. Remember, the capsule is gone in a second." Klaus looked at Dybo and said, "AJ, you know your job, but since this is the first time, and since it's so important, let's run through it one more time."

"Sure, Sir…"

"You ran the capsule into the launch bay." A nod. "The automatics attached the launch pouch and kick thruster." Another nod. "You moved it into the capture bay…" Nod. "…where it is suspended over the rail." A double nod. "So we're ready to go?"

"Yes, sir!"

"Okay, guys," Klaus grinned at Margo as he said that, "let's rock and roll!"

"The best view is over there," Margo said, pointing to the window looking west. She crossed the main deck, the others joining her.

"You guys suited up, Critz?" Klaus asked over his Link.

"Standing by here, Sir," Chrietzberg responded with a suited thumb turned upward.

A bright flash followed by a loud roar—and it was over. Within five seconds, the sun swallowed the dwindling thruster exhaust. "I'll check the capture bay just to make sure," Dybo said, donning his helmet.

※

Rodney Chrietzberg started his suit countdown timer as the capsule launched. He felt a certain degree of excitement, but it really was more a feeling of relief. He was a fifteen-year high-iron veteran, having worked every major high-rise project in the world during those years. But never, never had he done anything like this. No matter how he tried to visualize it, the *Amelia Earhart Skyport* simply did not seem like a tall building to him. He had done some ballooning down on the Jersey shore in his free time, and that was perhaps the most similar—but only slightly so. A balloon could be guided back to Earth—not this baby! When the capsule exploded, he figured they were done for. He and his crew had relived that moment a hundred times while waiting for this moment. He was beyond admiration for the designers of this incredible system. When the skyport righted itself, slowly, laboriously, he and the others watched in amazement. Physics in action like he had never seen it before.

"Okay, guys," he said to his crew, "I want you to spread around the deck. We tested this sucker to the limit once. Let's not give it another chance. When the capsule arrives, if something goes wrong, you guys use your high-iron balancing instincts to help the gyros stabilize this sucker. You got me?" High-fives around as the men positioned themselves to the best advantage. Chrietzberg placed himself at the window, looking east down the rail. He was determined to see the incoming capsule before it arrived.

This is the longest twenty-two minutes I ever spent, Chrietzberg thought as he stared out the window through his helmet. He glanced at the Earth below. He could see a thousand kilometers, although things got hazy near the horizon. It was too early to see the terminator, but he thought he could see twilight coming on—just beyond the horizon, so to speak. Baker was like a photograph directly below. Sunlight glinting on the westward marching waves gave the ocean surface a look that reminded him of the neon lights of Vegas. He checked his countdown timer—three minutes left. "Stand by," he warned his crew. He strained his eyes, focusing on the rail's vanishing point about a kilometer out. *Concentrate… concentrate!* And there it was, growing rapidly from a silvery dot to a disk, to a capsule nose…*It's coming pretty fast! Too fast! Oh shit!* He wasn't at all sure whether he had thought or shouted his last words…when the capsule slowed abruptly and glided into the launch bay. "Let's go service this baby," he said with obvious relief in his voice.

Pigman, despite his broken arm, and one of the other guys locked out and supervised the automatic removal of the spent kick thruster and launch pouch. The spent thruster ended up in a holding bin beneath the rail, and the launch pouch remained on the rail. The gantry moved the capsule to the dock. Once the capsule sealed to the dock, the men doffed their suits and stowed them. The next order of business was the hot meal waiting in the capsule. It wasn't exactly gourmet, but then it had been underway for a couple of hours already. The important thing—it was hot, and not candy bars.

※

Klaus left Shift Three alone to enjoy their meal, such as it was. While they ate, he and Margo discussed their next step, with Alex participating by Link. "I don't want them riding the return rail until we've tested it," Alex said. He got no argument.

"So, they're going to attach a fresh kick thruster to the capsule we just sent, turn it around, and send it along the rail, kick thruster first, braking with the kick thruster as it arrives here?" Margo asked.

"That about sums it up," Klaus said. "Our test protocol calls for several dry runs."

"But we're way beyond that," Margo said.

"I agree," Alex said. "Let's set it up."

"You guys finished stuffing your faces?" Klaus said by Link to Chrietzberg.

"We're ready for a shift change, Sir." A chuckle. "You gonna run an empty first?"

Klaus briefed him on their plan. "If it works without a hitch, you'll ride the next one back."

JARVIS ISLAND—JARVIS COMPOUND

Lori and Dex had managed to hitch a ride to Jarvis on the new floater that Carleton Montague and Delmer Woodward had shipped to the project. Lori's unfailing ability to excite the men and women around her charmed the new pilot, Noemi Wien, who found a legitimate reason to make the 1,850-kilometer trip. Lori hoped to convince her to remain until the rescue was completed, so she would have a guaranteed trip back to Baker, and so they would have time to get to know each other better.

While Chrietzberg and his men completed their first real meal in several days, Lori collected whatever background material she could, and Dex roamed the island, recording stock footage to fill the background of her coverage. Lori placed a call to *Earhart Skyport*. "Critz, it's Lori. How're you guys doing?" She paused. "I know you guys are busy as all get-out, but do you have a couple of minutes to share your thoughts with my audience?" She left unspoken the implied completion of the sentence—*in case you don't make it.*

"Glad to, Lori. We have a couple of minutes. You can use my video pickup."

"I understand you just completed a hot meal—the first in a while?"

"Yep…and it sure beats the candy bars we've been eating since the explosion." Laughter all around; it was obvious the men were feeling good about the meal, but apprehensive about the next step.

"Are you at all nervous about the coming capsule trip?"

"Nervous? We're high-iron men. We don't get nervous."

Lori thought differently, but she had no desire to pop any bubbles. "So, what's next?"

Chrietzberg told her about the test, and explained that they would depart in the next arriving capsule. "We should be on *Noonan* in less than an hour, and on Jarvis a few minutes after that."

"Can I talk with you then on camera?"

"Sure thing."

Lori signed off and grabbed Dex to get some shots of her by the launch facility. Noemi Wien sauntered up to watch, her blue-gray eyes hidden by mirrored pilot's shades. Her pixie-cut auburn hair glistened in the afternoon sun. Khaki short shorts and matching tank top covered her tight, small-breasted body.

"Did you know it sometimes gets this hot in the Alaskan bush…for a day or two, anyway?" she said to Lori. "I got my wings up there. Followed the path of my great-great-whatever-granddaddy, Noel Wien."

"Who?" Lori asked.

"Noel Wien—founded Wien Air Alaska in the 1920s."

"Really…"

"Yep. He used to fly hunters into the bush. One time he landed on a short beach, and had to beat more pitch into his propeller with a sledgehammer before he could take off. He's my hero."

Lori grinned. "After the rescue, Noe, we need to get together for an in-depth interview. I know my audience will love your story. You okay with that?"

"Love to," Noe said, hands on hips, legs apart.

Although Lori couldn't read her eyes through the pilot's shades, she was certain Noe had sent her a subliminal message as well.

FRED NOONAN SKYPORT

Chrietzberg and crew attached a fresh kick thruster to the capsule, pointed the business end east, reattached the launch pouch, and readied it for launch down the rail. Klaus gave the signal, and the launch pouch engaged. This time Chrietzberg watched the capsule nose fade into a point that he quickly lost in the darkening, star-filled firmament. Since the test seemed to be working, he allowed his crew to congregate around the window overlooking the rail—the one with the view. They could now distinctly see the terminator sweep over the horizon to continue its inexorable march toward them.

About twenty-two minutes later, Dybo commenced a running dialog with Chrietzberg as he readied the capsule for another run to *Earhart*. It seemed a shorter wait this time as the arriving capsule glided to a halt in the bay. *Only the second time, and it's already becoming routine.* "Let's go, boys. Get her ready. We've got an appointment with groundside."

A half-hour later, after checking and rechecking, Chrietzberg was satisfied that nothing had been left undone. He and his six companions found their seats in the capsule, seats that had been rotated front to back for the reverse launch; he sealed it, and let the automatics take over.

The capsule dock hatch slid down and sealed the lock. Then the hydraulic rams eased their pressure, and the capsule settled into its gantry harness. With a slight jerk, the gantry moved the capsule down the track and across to the rail. The capsule moved back against the kick thruster, where a clank signaled the successful attachment of the thruster. Then the gantry performed the half-turn, and the capsule shuddered slightly as the clamp attached the launch pouch. A disembodied voice commenced a short countdown.

Five…four……one…zero…

Chrietzberg felt himself pushed back into his seat with a very tolerable amount of g-force. It felt very much like stepping hard on the accelerator of a car and keeping it there for a while—eleven minutes, to be exact. Then a soft chime signaled the seat rotation just before the thruster ignited. Chrietzberg began to feel a bit uneasy, but the feeling vanished with the sound of the thruster. Almost immediately, he began to feel the deceleration as he was pushed back into his seat back again. As before, the feeling was not uncomfortable. Before he had a chance to think much about the process, the deceleration lifted as the capsule glided into the bay at *Fred Noonan Skyport*. Several clanks signaled removal of spent thruster and the launch pouch, and then the gantry moved the capsule to the bay where the hydraulic rams created a seal with the lock.

"Welcome back!" Klaus said as the gull-wings opened. "I figured you guys would want to take a few minutes before continuing to groundside."

"*Kwe AJ, nia:wen!*"

Klaus glanced at Dybo. "He says Hi and Thanks!" Dybo said with a grin.

"*Ianeratie' ken?*" Prajeet Kahnawak asked.

"*Tiohrhen:sa sata:ti*," Dybo told them. "Say it in English, Jeeter."

"Sure, Boss. I asked, 'How're you guys doing?'"

"Thanks, all of you. We're doing great…now!" Chrietzberg said. "So, when does the next capsule depart?"

"Where? Back or down?" Kahnawak asked with a wide grin.

"Down, Baby, down! Let's get the fuck outa here."

CHAPTER TWENTY-NINE

SEATTLE—SMITH TOWER

Mabel Fitzwinters took John Boyles' encrypted call at her Smith Tower desk, where she seemed to be spending most of her time lately. She had followed the rescue of Rodney Chrietzberg and his crew with wonder and admiration. How had she been so lucky to get these giants on one team, working together, and actually enjoying it? She settled her bulk back in her chair to hear Boyles out—her intuition telling her this would be a long one. "John," she said. "What's up, my old friend?"

"I think we have a line on the bad guys," Boyles told her. "You and I both know that the DPRK is behind the whole thing. Well…some of the pieces just fell into place. My guys are chasing the money trail right now, and it points directly at General Jon Yong-nam." Boyles then proceeded to brief Mabel on the details of their legal investigation. "We're pushing the legal envelope on this one," he finished. "But it's airtight."

While they talked, Mabel's display announced an incoming call from Rex Johnson. "Rex, I'm speaking with John. Let me conference the three of us together." Johnson and Boyles exchanged greetings.

Mabel then briefed Boyles and Johnson on the status of the repair effort. The cables had been manufactured and were on their way to Baker, to arrive in three days. While they were waiting, Klaus and Margo tested all the various modes for capsule launch, both away from Earth and between skyports. Simply stated, they found no real problems. The hardware and software did what they were supposed to do. The only remaining test, she told them, was a launch abort to the atmosphere. "Alex wants to hold off on the abort test until after the skytower replacement. They're commencing that process tomorrow."

"My reason for calling," Johnson said, "is to give you the latest developments in Congress. After the sabotage, somebody in the Transportation Committee suggested federal oversight of Slingshot."

"Who?" Mabel shot at him.

"Don't know," Johnson answered. "These things often happen in the dark of night, so to speak, and come to light as full-blown undertakings. This matter came up in passing, my contact tells me, and was assigned to the Highway Safety Subcommittee. That subcommittee has a meeting this afternoon where I expect to brief the members on what actually happened."

"Rex," Mabel said, "the last thing we want is for government inspectors to second guess everything we do."

"I agree, Mabel, but the cat is out of the bag on this one. We can fight it, but we'll lose. Far better that we control the process, limit inspections, and keep them at arm's length. I'll know more following the meeting this afternoon." He smiled ruefully. "We're not alone, you know. Many members still try to limit federal growth."

"John," Mabel said, "do you have anything to add?"

"Not yet, Mabel," Boyles said, "but I think you need me with you this afternoon, Rex."

"I agree," Johnson said. "I'll set it up."

"You keep a tight rein on this one, Rex," Mabel said. "I don't want this blowing up in our faces."

JARVIS ISLAND—JARVIS COMPOUND

Alex was feeling pretty good about the state of things. He continued to wear a tape, but the pain was nearly gone. Apryl still insisted on nursing him, but even she could see that he was virtually healed. When Noe arrived with Lori and Dex, Sanchez returned to his regular duties, ferrying men and materials in the general Eastern Complex-Jarvis Island area. Lori was present for every important event, but Alex noticed that when nothing was happening, both she and Noe were nowhere to be found. Alex chuckled to himself. At least that problem was no longer his, for the time being anyway.

At lunch, he sat with Margo, talking about where things were going. "We need to get back to Baker," he told her. The new cable was arriving in a couple of days, and they needed to be ready. "We have to commence descent, so that by the time *Earhart Skyport* reaches nine klicks, we'll be ready to attach the new skytower."

The Baker process was multi-stage. First, haul a pilot line up nine kilometers with the Chinook, and attach it to the terminus under the skyport. This meant, of course, putting a team back on *Earhart*, rigging a temporary platform, and working out the details for hanging a man over the side to capture the pilot line from the Chinook. Then, using the pilot line, pull up the skytower cable and another pilot line, and attach the skytower to the terminus. Following that, use the second pilot line to haul up the deflector that would, itself, carry a third pilot line. Finally, pull the lift cable through the deflector with the third pilot line. From that point, it was merely a matter of continuing the rail ascent while feeding both skytower and lift cables from barges moored in Meyerton Harbor. Alex had already arranged for the reinstallation on Jarvis of the A-frames to accommodate the total cable bundle, and on Baker for the suspensor/lift cable combination, and another set of roller-covered A-frames for the skytower cable.

Back on Jarvis, they had to shut down the boost and lift cables, and as *Noonan* descended stow seventy-one kilometers of cable in a way that it could safely and conveniently return skyward when the rail started its ascent immediately following the final hook-up at Baker. The plan was to moor three ocean-going barges in tandem along the inside of the Millersville breakwater, set to receive the skytower bundle as it descended, serviced by a lumbering Sikorsky Skycrane. Klaus had calculated that the Skycrane would just be able to carry the cable load at an altitude of fifty meters or so, which was more than sufficient to fake it down across the three barge lengths.

Alex called up the four high-iron shift supervisors currently on Jarvis and explained his need for two volunteers to return to *Earhart*

Skyport for the duration of the reinstallation of the *Baker Skytower*. "You'll go up this evening," he said, "with sufficient food and water for several days. As soon as the receiving barges are in place here at Jarvis, we'll commence the descent. You'll rig a platform at the Baker terminus, which should keep you occupied for a couple of days. As soon as the skyport reaches nine klicks, we'll send up the Chinook with the skytower pilot line." He then outlined the entire procedure so his volunteers would understand what they were getting themselves into. "You'll be the focus of international attention as the world follows your daring," he said, and added as an afterthought, "As a precaution, you'll wear chutes." Alex paused to let the information sink in, and then said, "Give me two names in the next few minutes."

Alex called Noe, and told her to be ready in an hour to fly Margo and himself along with Lori and Dex back to Baker. He asked Klaus to supervise the activities at Jarvis.

BAKER ISLAND—BAKER COMPOUND

Lars Watson sat in a simple wooden chair resting his head in his arms on a plain table in a locked room in the Baker Complex. He wore dirty shorts and a torn T-shirt, feet shod in ragged sneakers without laces. His still short hair was unkempt, and he needed a shave. He was discouraged and afraid. He had done his job—flawlessly. Had nearly escaped. He cursed his luck. He did not know the status of Slingshot, but something from the sky had destroyed the skimmer. It was over for him; that much was obvious. People had died, and they had him locked up. The future seemed pretty obvious. The only thing he could do now was use the trial setting as a platform for getting the word out.

He missed his Link. Without it, he felt incomplete. Only once since his teens had he not worn the ubiquitous device—when he was apprehended diving the buoy. It was in a very real sense an extension of himself. He felt naked and alone. The door opened, and someone he didn't know set a tray of food on the table. Watson swept it to the floor, shouting, "Leave me the fuck alone!"

"You need a shower, you animal," Noah growled as he left the room, leaving Watson's mess on the floor.

Watson winced at the insult and stood up, marveling at how much his body hurt as he did so. He remembered being captured, seeing the explosion on *Earhart Skyport*, and Alex hitting him. The next thing he remembered was lying on the deck, restrained with duct tape, and then the skimmer was cut in half by something from the sky—something from Slingshot. Next thing he remembered was a couple of dolphins bobbing him to the surface and keeping him there, face up so he could breathe. He shuffled to the shower to scrub away the grime and filth. The fates had kept him alive, so he still had something important to do. He stepped out of the shower dressed once again in the grimy shorts and shirt, but with head held high. He was a hero, and one day the entire world would know.

The door opened again, and Noah ordered him to follow. Watson trailed the Maori steward down the hall to Alex's Baker office. For a moment, he considered bolting but quickly realized that if he did, he might not make it back alive. These people were pretty pissed at him.

He walked through the door into a three-way conversation between Alex and two men in suits. One of the men was signing a document on Alex's desk. The other turned to him, gripped his shoulder, spinning him around, and expertly shackled his wrists behind his back with handcuffs. Almost immediately, the men ushered him from the office, out the building door, and into a pickup. Five minutes later, the pickup pulled alongside the Chinook. They lifted into the tropical morning sky as birds scattered from the spinning tandem blades. Twenty minutes later, following an uncomfortable flight with his hands still shackled behind him, Watson stepped out onto the Howland tarmac right at a stairway leading onto a private jet. One of the men hustled him up the steps and into the aircraft. He removed the handcuffs and ordered Watson to sit and buckle himself in. Watson started to ask their destination.

"Shut up!" the man told him, "Or I'll tape your mouth shut."

Shortly thereafter, the other man entered the cabin, and five minutes later, the jet was airborne. Watson decided to keep a low profile during the flight. A male flight attendant appeared shortly after takeoff offering something to drink, but when Watson glanced at the man who had hustled him aboard, he decided to refuse. He still was dressed in dirty shorts, torn tee-shirt, and laceless sneakers.

Three-quarters of an hour into the flight, he began to feel cold, but the hostility he felt from both men kept him from requesting a blanket. To his total surprise, fifteen minutes later, the aircraft began to descend.

One hour and fifteen minutes after takeoff, the jet rolled to a stop, and the man who seemed to be directly in charge of him ordered him to stand up. Watson stood, and in anticipation of being handcuffed again, put his hands behind his back and turned around. Instead, the man ordered Watson to face him. He handed Watson a packet of papers and a link, and escorted him down the stairway to the tarmac. Then without a word, he turned, reentered the plane, and closed the door. Watson looked around, squinting in an obviously tropical sun. He glanced at the papers he had been given—they were identification and travel papers, and some cash. He walked toward the passenger terminal as a refueling truck rolled up to the jet. An official met him at the entrance and accepted his papers.

"Welcome to Pago Pago, Mr. Flannigan."

JARVIS ISLAND—JARVIS SOCKET

Klaus joined Ajay Dybo and Rodney Chrietzberg in the capsule that was awaiting launch to *Fred Noonan Skyport*. Lori spoke briefly on camera with all three, explaining to her audience that the two Mohawks had volunteered to return to *Amelia Earhart Skyport* in preparation for lowering the entire loop to a mere nine kilometers to attach a new skytower to *Earhart Skyport*. Alex and Margo stood to the side, observing the proceedings, but remaining in the background, since this was mainly a Dybo-Chrietzberg show. Klaus appreciated the way Alex made sure that people doing a job received the credit. He didn't need any credit for himself, but he suspected that the worldwide attention AJ and Critz were getting at this moment would be one of the highlights of their lives.

"Are we ready, guys?" Klaus asked, getting silent thumbs-up from both men. He nodded at the launch operator. The gull wings closed, the capsule moved into launch position, pivoted to vertical, and they were off.

At the skyport, the shift crew parked the capsule in preparation for the trip to *Earhart*, while a freight capsule ascended with food for a week and platform construction materials. The food was mostly freeze-dried—not exactly gourmet, but it didn't taste bad, and it was nutritious. Klaus had to admit to himself that he was glad he was not going with the two. He was not particularly concerned about their safety, although they really were taking a calculated risk. His concern was the boredom. He figured they would complete the platform in no more than two days. It would take several more to reach nine klicks. That was the period that he would not enjoy.

Klaus watched the shift crew prepare for the transit to *Earhart*. The process seemed almost routine. The kick thruster flash announced the capsule's departure, and the shift crew got the freight capsule ready to follow. Klaus was not yet ready to run two close transits on the rail. He still insisted on one at a time. When Dybo announced their safe arrival and reported that the capsule was parked, Klaus told the shift boss to launch the freight capsule. During its transit to Earhart, another passenger capsule arrived from Jarvis. Once the freight capsule was safely stowed at *Earhart*, Klaus called Alex.

"We're done here. We'll be back on Jarvis in a few minutes."

"I'm airborne with Margo, Lori, Apryl, and Dex," Alex told him.

"Is Noe piloting?" Klaus asked. Alex said she was. "Kinda outnumbered there," Klaus said with a chuckle.

Klaus and the shift crew boarded the waiting capsule and descended to Jarvis.

AMERICAN SAMOA—PAGO PAGO

Watson took a cab to the center of Pago Pago, and walked to the clothing store he had used during his last visit, where he purchased a suitable set of clothing, letting the clerk choose the colors. He checked into a local hotel as Ed Flannigan. He showered, shaved, and—dressed in his new clothing—went to the open-air restaurant from his last visit and ordered a tropical cooler. While sitting at the table, sipping his drink, he reviewed the papers he had received as he arrived in Pago Pago. He was Ed Flannigan, a tourist on a short vacation in Pago Pago. The papers also supplied him with encrypted access through a public Link. Using the restaurant Link, he received another encrypted connection that would work from his new personal Link. So his benefactors still wanted complete anonymity. That was fine with him. Somehow, they had sprung him from the feds, given him money, and still seemed to want his services. *So long as I get what I want, they can have what they want*, Watson thought as he returned to his room to make the encrypted call.

Watson established the encrypted link, but instead of a face, he got a general background. He was given account numbers for accounts against which he could draw, but was told to keep his expenses low—to maintain a low profile. "Grow a beard and keep it trimmed short, dye your hair dark, and wear brown contacts," the Link instructed. "Use the ticket in your packet and travel to Los Angeles as tourist Ed Flannigan. Upon arrival, check into the Airport Marriott as Neil Lansing, an inspector with the Federal Highway Administration. Destroy the Flannigan papers, and wait for further instructions."

BAKER ISLAND—OPERATIONS CENTER

Alex paced back and forth in the Baker Operations Center, waiting for all the elements to be completed before commencing the descent. Margo coordinated the waterborne activities on both islands and the Western and Eastern Complexes. Alex assigned Klaus to focus on the Jarvis operation since it was critical that the descending skytower and lift cables be handled without damage, and in a way that they would be ready for immediate return to the sky. One corner of Alex's consolidated holodisplay showed regional weather patterns. The descent and ascent could be accomplished in virtually any weather, but the skytower hookup nine kilometers over Baker was tricky at best, and he didn't want to contend with any weather complications. Rather than risk it, he was prepared to hold the process to let a weather system pass. A weather system seemed to be developing about halfway between Jarvis and Baker, but it was unstructured and didn't appear to be an immediate problem.

"May I steal you for a short interview?" Lori asked from over his shoulder. "My audience is prepped for what comes next, but I know everyone would love to hear you tell them once more what they will see in the next several hours."

Alex was keenly aware of the value that positive publicity brought to his operation. Since Lori was behind the mike, he already knew the message would be supportive—so long as he kept up his end and supplied positive material. He turned to face Lori, focusing on Dex's holocams just over her shoulder.

"Alex Regent—the man of the hour," Lori said to her audience across America and her growing audience around the world. "We are speaking with the Boss, the man who, more than any other, has been the power behind this incredible project. All of my regular viewers know what happened a few days ago. You saw the massive pile of skytower cable, the crushed buildings, the shattered hopes and dreams of the men who died, and you experienced the fear of the men stranded in the skyport. But you also watched the cable pile disappear that same morning, and you witnessed the triumphant return of seven heroic high-iron men—all guided by this man, my friend, Alex Regent." She smiled warmly at Alex, and stepped close to bring them together on-camera. "What's happening now, Alex?" she asked brightly.

"Two volunteers just returned to *Amelia Earhart Skyport*—Ajay Dybo and Rodney Chrietzberg, two of our highly competent Mohawk shift leaders. The extraordinary thing is that Mr. Chrietzberg had just returned from several days stranded up there, not knowing whether he would ever see the ground again." Alex indicated the holodisplay covering the wall behind him. "This is what they intend to do." He zoomed to a close-up simulation of the terminus, where two space-suited figures constructed a platform that would enable them to attach the end of the skytower pilot line when it arrived by Chinook in several days. "The key is that the Chinook has a ceiling of about nine kilometers, so we need to lower Slingshot from its present eighty-kilometer height all the way down to nine." As he talked, the holodisplay showed the rail descending, and followed the progress of the Chinook as it hauled the skytower elements skyward.

From behind Lori, Margo waved to Alex, indicating that something needed his attention. He smiled at Lori and her enthusiastic audience, and excused himself. "Gotta go…duty calls," he said to the cameras.

"We're ready to commence descent," Margo told him. "Klaus has his barges and the Skycrane in place. I'm monitoring all six of the rail tensioners and the twenty downslope tensioners. I got individual techs watching each tensioner. We're just awaiting your starting order."

AMELIA EARHART SKYPORT

"Well, cousin," Dybo said to Chrietzberg, "it's you and me, buddy!" The two Mohawks exchanged a high-five.

"Let's fix this puppy, AJ!"

The two men pressurized the capsule dock and then opened the large door into the main skyport. They unloaded the platform parts from the capsule onto the capsule dock. Dybo closed and sealed the hatch leading to the main deck, while Chrietzberg sealed the capsule hatch. They both donned suits and helmets, and then Dybo started the evacuation pump to sweep the air from the chamber, compressing it in bottles for reuse. It took the better part of five minutes, but finally, the flashing yellow caution light switched to solid red, and Chrietzberg opened the capsule hatch, releasing the remaining air with a slight whoosh. Then they inflated each piece, and turned on the UV floods in the chamber to cure the pieces to hardness.

The suited men stuffed tubes of adhesive into their leg pouches, and each picked up a platform piece. Together they trudged along the track out to the terminus. Immediately below their feet was eighty kilometers of nothing, not even the tenuous thread of a skytower disappearing into the emptiness below. From the track, the illusion of being in a high balloon was overwhelming, and Dybo had to reset his perception. His high-iron sense demanded an accurate perception of every situation he found himself in, including this one.

The assembly proceeded like an erector set project, except they stuck each inflated piece in place with vacuum-cured adhesive, and the pieces were measured in meters or decimeters instead of centimeters. By late afternoon, they had completed two levels of a three-level platform extending ten meters below the terminus. They shut down the construction for the night rather than take the time to mount floodlights to work another few hours in the dark. They returned to the capsule dock but did not repressurize it. Instead, they locked onto the main deck through the lock, stripped out of their suits, and retired for the night.

At first light, Dybo rolled out of his rack and mustered Chrietzberg to his feet. They grabbed a quick breakfast, washed down with black coffee, suited up, and locked back into the capsule dock. Ten minutes later, Dybo was handing platform sections down to

Chrietzberg, who balanced on a rigid section forming the base of the lowest level, as he glued the remaining pieces in place.

With the platform completed, the men adapted the gantry winch to serve as the up-haul winch for the operation. A pulley at the terminus guided the cable to the winch. It was a simple arrangement that took advantage of what was already in place.

Four hours later, Dybo called Alex. "We're done up here, Boss, at least until the Chinook arrives. How soon will you commence the descent?"

"Commence?" Alex chuckled. "You've been descending for the past six hours!"

CHAPTER THIRTY

SEATTLE—SMITH TOWER

Mabel Fitzwinters stood at her south-facing windows sipping on a full-bodied Kona coffee. The rising sun off to her left glinted off the distant snow-covered slopes of Mt. St. Helens. To her right ungainly orange cranes along the wharf already moved in a ghostly pantomime, loading containers on large container ships for destinations throughout the Pacific. She let her eyes cross Puget Sound, lingering on two ferries painting white wakes in opposite directions on the still water. Beyond that, land piled on land drawing her eyes northward to the snow-capped Olympic Mountains, appearing as if a giant fist had pushed through the Earth's crust, tilting huge slabs away in all directions. Mabel walked back to her ornate desk to receive a Link call from John Boyles on her newly installed, layer-encrypted Link system.

"Mabel…"

"Good morning, John. I've been awaiting your call."

"The good news is the Highway Safety Subcommittee didn't hit us too hard. We're subject to an annual inspection at each skyport, the launch facilities, and both the OTEC generators and linear accelerators. Since

Slingshot is entirely new in concept, our engineers will work with the FHWA inspectors to develop a reasonable inspection protocol. It's our opportunity to keep it simple and straightforward."

"So, what's the bad news," Mabel asked.

"They want to inspect *Earhart Skyport* before any commercial operation."

"What can a government inspector possibly know that we don't?" Mabel asked, the frustration rising in her voice. "We've built a brand new industry, created thousands of jobs, and set the stage for inexpensive, general space transportation, and these idiots want to regulate us?" Mabel was fuming. "Just who the hell do they think they are?"

"Unfortunately," Boyles said with a conciliatory smile, "they're the 800-pound gorilla in the room. So long as our operations touch American soil or terminate in American waters, they will claim oversight. We can fight them in court, but we'll lose eventually, and they'll hold it against us. Better we cooperate from the start, Mabel."

"Any update on the DPRK, John?"

"Not DPRK so much as *Environment, Inc.*," he said. "General Jon Yong-nam pulls the strings, I'm certain of it, but he's untouchable. We can't find a chink in his armor. We're closing in on *EI*, however. These guys are clever beyond belief, but we've isolated and decrypted a string of Link comms that clearly shows their operational tactics. We can now prove they funded Watson's activities right up through the *Aku Aku* operations. Once we have hard evidence of their control over the *Earhart* explosion, we'll turn it over to the federal prosecutor. His office will issue the warrants, and we'll shut them down permanently." Boyles paused to let the information sink in. "We're nearly there, Mabel. We're going to nail those bastards!"

HONOLULU—AIRPORT MARRIOTT

Neil Lansing, a.k.a. Lars Watson, checked into the Marriott front desk from his Honolulu flight continuation from Pago Pago. He appeared tanned and well rested, but was beginning to show a beard that had missed several days of shaving. His room was paid in full for two days, and was located on the far side of the hotel, so he would be able to leave the hotel without passing the front desk. As he turned to go to his room, the counter clerk handed him a package.

"Here, Mr. Lansing, this arrived for you this morning."

In his room, he opened the packet. It contained hair dye, beard and mustache dye, permanent contacts, identification papers, and travel documents to Seattle, leaving in two days. His passport photo showed a face he didn't recognize, a bearded countenance with medium-length dark brown hair, according to the description, although it looked medium-dark gray to Watson. Beard and hair color matched, and the eyes were brown, again, according to the description. The papers included an encrypted Link connection, with instructions to make contact at five pm. Watson glanced at the time; it was only three. With a couple of hours to kill, Watson showered and dyed his hair and beard. Since his beard was so short, he retained the beard dye for a second application before leaving in two days. He inserted the contacts and examined himself in the mirror. A stranger looked back at him. *Amazing*, he thought, *a few simple changes, and I am an anonymous traveler with no connections to anything Pacific.*

Using extra caution, Watson bundled the leavings from his makeover, and stuffed them in a waste container by the ice machine on the floor below. His room was equipped with a standard business shredder, so he made a mental note to shred his Flannigan documents before leaving.

At five pm, Watson activated his Link and placed the encrypted call. As before, he saw no face, just an impersonal background. "When you arrive at SeaTac," the disconnected voice told him, "look for a greeter holding a Lakeview sign. He will hand you an envelope. Take it and leave the airport for downtown Seattle by public transportation. Find a hotel—they're all over the place. At four-thirty-four, exactly, present yourself at the *Seattle's Only Java* cart in the main lobby of the Smith Tower. Tell the cart operator you want a *Java with double cane*. Use those exact words. Carry the coffee back to your room. Don't drink it!"

The voice then went on to explain the contents of the two vials in the coffee, and how they were to be used. It explained that since the explosion, all suits were equipped with specially designed parachutes that opened automatically at the proper height. After he placed the gel, he was to jump and let the automatic system take over. A vessel would locate his beacon and pick him up. The voice told him to report to the main offices of *LLI* the following morning. They were expecting him, and would extend to him full cooperation. They would fly him to Howland, and would put him up at Baker until they were ready for his on-site inspection of the skyport, which would be within a couple of days of his arrival.

"Remember," the voice told him, "you're a government inspector. Stay aloof. Don't make any friends. Don't talk with anyone except directly in connection with your inspection. And place the gel at the very first opportunity, and as soon as you place it, jump. Your suit will take over from there."

BAKER ISLAND

Klaus had his hands full as the rail continued its inexorable descent. Shortly before the descent commenced, he had de-energized the lift cable, using the boost cable to brake it. Then he de-energized the boost cable, transferring some of its energy back to the lift cable, so that over a couple of hours, the entire system came to a halt. Then his team disconnected the skytower cable bundle from the socket. In the meantime, the Sikorsky Skycrane got underway, and flew to a parking position, hovering as close as possible to the skytower. When the skytower cable bundle was fully disconnected from the socket, the Skycrane lowered a special harness that the crew attached to the end of the bundle. The Skycrane lifted the end of the bundle into the air, and as the rail high above descended, the Skycrane carried the bundle over to the barges while keeping the catenary twenty to thirty meters off the ground. The Skycrane pilot carried the end of the bundle past the barges and the breakwater for a full three barge lengths. Then he laid the catenary on one end of the linked barges, and flew back along the barges laying the cable along their length. As soon as he placed the end of the bundle on the third barge, he flew back to the socket to pick up another loop and carry it out to the barges…out and back, out and back—seventy-one kilometers' worth.

Each section was heavy, so much so that the Skycrane could only rise to about fifty meters. The pilot's job was to keep the bundle out of the dirt, and, if possible, out of the water as well. It was tricky, but Klaus had some control over the descent rate by slowing or speeding up the ribbon. Every couple of hours, the Skycrane needed refueling. Although the actual refueling took only fifteen minutes, the entire fueling operation was closer to thirty minutes. For this, Klaus had to halt the descent, but the tensioners were holding tight, and the ribbon remained stable during the stops.

During each refueling, the maintenance crew that arrived with the Skycrane ran a different diagnostic, ensuring nothing went wrong. Bobby Pfaff hung around the maintenance crew, soaking up every bit of knowledge he could. Klaus finally had to send him to his quarters for a sleep break. When Klaus told Alex about Bobby's enthusiastic participation, Alex told him, "You need a sleep break yourself, Klaus. Take a couple of hours. I'll spell you from here. If things get out of hand, I'll stop the descent while we straighten it out."

Reluctantly, Klaus agreed. Although he could have gone on a while longer, he knew that without sleep, he would not be at his peak. He officially turned things over to Alex and crashed on a cot in the back of the Jarvis Operations Center.

It seemed like no more than a few minutes when a team member shook him awake. "It's been three hours, Sir. The Boss is calling."

"Yeah, Alex, what is it?"

"I need you to spell me for a couple of hours," Alex said, peering from the holodisplay with bleary eyes.

Klaus shifted half of the holodisplay covering the Control Center wall to Alex's display. "I got it, Alex," he said, sipping on a scalding cup of coffee somebody handed him. "Get some sleep, friend." He surveyed the Baker display, noticing the storm picture at one end. It still was well to the east of Baker, but it was gaining momentum, and the winds were picking up.

BAKER ISLAND—CONTROL CENTER

Three hours later Alex stumbled into the Baker Control Center. The ribbon was at twenty kilometers—eleven more to go. He checked the storm. It was definitely on a direct path toward Baker—twelve to eighteen hours away. *That sucks!* He thought. *We need to attach the skytower right about then…shit!*

"Klaus," Alex said over their Link, "can you speed up the descent?"

"What's up?" Klaus asked.

"I'm looking at that storm…"

"Yeah, me too," Klaus said. "What's the timeline?"

"Twelve to eighteen hours," Alex told him.

"That's not good." Klaus paused, his face thoughtful. "We're getting pretty good with the Sikorsky. We have refueling stops down to just under twenty minutes. I think I can give you four, maybe five hours."

"If the storm stays on track, that just might be enough," Alex said. "Let's do it…and, thanks."

Alex pondered the situation. Normally, equatorial storms started near the equator, moving west, and tracked off either to the north or south. Only rarely did they stay near the equator. But this one had formed west of Jarvis, and stayed right on track for Baker, growing as it traveled. It was now about 500 kilometers out, moving at about thirty kilometers per hour, and it packed winds of over sixty knots. Unless something changed, it would cross Baker in sixteen hours or so, with winds in the eighty-knot range. They would begin to feel the effects in about five hours.

The storm system was 150 kilometers across and three, maybe four kilometers high. It didn't have a lot of structure, but the winds were anti-cyclonic, and it contained a poorly formed center about twenty kilometers across that was relatively calm. The Chinook could handle such weather, but Alex's problem was that the action would be taking place five to six kilometers above the top of the storm, and the chopper would be hauling the pilot line up through the storm. Hovering at nine kilometers was virtually at the upper hover limit for a modern Chinook, stripped of everything but what was necessary for the job. In this case, however, the hookup would take place right in the middle of the high-level trade winds that blew a steady 60 to 80 knots east to west. This meant an easier hover resulting in greater flight time. Normally, a Chinook carried about six hours of fuel, but at nine kilometers altitude, engine efficiency dropped way off as fuel consumption skyrocketed. He figured they had a maximum of thirty minutes to hand off and secure the pilot line.

The storm appeared to be tracking directly over Baker, bringing the eye across the island, with a window of relative calm lasting about an hour. Logistically, it was easier to feed the pilot line from the Chinook than to haul it up from the ground. The Chinook carried a reel that held the entire nine kilometers of line that would unreel as the chopper ascended. The upper end of the pilot line was attached to a lightweight 300-meter line with a small chute at its end. The large reel was designed to detach a face and dump its entire contents into the air on order.

With about eight hours before he sent the Chinook into the sky, Alex ordered Klaus to take a four-hour sleep break. He assumed control of the Jarvis operation while Klaus rested, but a scant three hours later, Klaus reported back to duty, sipping from a steaming cup of coffee. When Alex raised an eyebrow, Klaus grinned. "Yep...one cream, two sugars...as always." Then he added, "Get some rest yourself, Alex. We'll both need it in the next few hours."

Thankfully, Alex dropped to the cot behind him and was instantly asleep.

Minutes later, it seemed to Alex, Margo shook him awake. He could hear the wind howling outside and the rain drumming fiercely against the windows. She handed him a black coffee. "It's stronger than usual," she said. "You're gonna need it."

"Thanks, Doll. Is everything ready at your end?" He knew it was, but he asked the question anyway.

Margo nodded. "We even have boat crews standing by in case something happens to the Chinook," she told him. "Bad break, this storm, but we're as ready as can be."

Alex looked at the storm display that now dominated his holo-display. He checked the time. "It's tracking true. We've got about an hour before we send up the Chinook. I'm going to check the setup personally."

"Alex...No! The wind's way too strong. You can't do anything the guys haven't already done. Please..."

Alex walked over to the outside door and cracked it open. A blast of wind and rain entered through the crack, and it took all his strength to shut the door again. He grinned at Margo. "I guess you're right." He shook water from his hair and face. "I guess I have to rely on the chopper crew." He checked the time again—almost six AM. The sun was rising, but it was not apparent from Baker. Heavy rain clouds blotted it out completely. As the hour passed, however, the sky brightened and the winds lessened. Then the rain stopped and the winds died away.

"Get the chopper underway," Alex ordered on All-Call. He stepped outside to watch under a mostly blue sky absolutely swarming with skreeghing birds. To the east, the ocean was open, filled with choppy waves. To the north, south, and west, however, walls of dark clouds blocked the view. The distinctive whine and characteristic syncopated woop-woop sound of the Chinook filled the air as the big bird rose into the morning sky, trailing the pilot line from its side door. Alex checked the time again as he watched it rapidly shrink with height until he could no longer see it.

"The Chinook will be there in twenty minutes," Alex announced to Dybo and Chrietzberg, high above.

AMELIA EARHART SKYPORT

Dybo and Chrietzberg had not slept the past few hours. There was not a lot to do, but they discovered to their considerable surprise that as the skyport lowered into the atmosphere, things got pretty noisy. Dybo had not given it much thought before, but on reflection, it seemed pretty obvious. Intense equatorial sunshine heated the surface air, causing it to rise. This rising air was replaced by air moving along the surface toward the equator. Because the Earth rotated toward the east, the air flowing toward the equator on the surface turned west, forming the trade winds. At about twelve kilometers or so, the rising air moved away from the equator, turning eastward due to the Earth's rotation. It got pretty complicated after that, but it was pretty obvious that the skyport was being buffeted by some strong winds out of the west. Fortunately, the air was thin enough that the structure could withstand the generated forces. It was noisy as hell, though, as the wind vibrated structural elements in passing, creating a cacophony of sounds that sounded mostly like a symphony orchestra tuning up before a concert.

They were already suited up when Alex announced the departure of the Chinook. "Let's go do this, cousin," Dybo said, giving his friend a high five as they locked onto the capsule dock.

Dybo took the lead, ending up at the bottom of the ten-meter platform hanging from the terminus. Chrietzberg placed himself a couple meters above Dybo. They both clipped themselves to the framework and awaited the Chinook. The wind was easily 60-knots, perhaps a bit more. It buffeted them fiercely, pressing them into the framework. Dybo looked down at the storm. He could see the entire thing, and the clear water and sky beyond its limits. Directly below, the twenty-kilometer-wide ill-defined eye revealed a clear view of Baker, near its western edge. The rest of the storm was a roiling mass of surprisingly white clouds, illuminated by the rising sun. He strained his eyes and saw a glint directly over the island center.

"I see it, Critz," Dybo said, pointing down.

"Where?"

"Right there." Pointing.

"You're blocking the view!"

Dybo was surprised at how rapidly the big helicopter took shape. In three minutes, he could clearly see the trailing pilot line with a huge catenary billowing to the east. The chopper pilot waved at him and carefully stationed the bird about 300-meters to the west with the open door facing the skyport. A suited figure in the door waved at them, and then began feeding the lightweight line out the door. Immediately the wind caught the chute as the cable paid out to its full length. The air-filled chute seemed to float about thirty meters down and some fifty meters to the east.

Dybo resisted the temptation to shout instructions to the pilot. The chute and the bottom of the platform carried precise locator beacons that the pilot used to manipulate the chute to Dybo. As Dybo watched, the chute began to inch its way west, until two minutes later, it was thirty meters directly below him. Dybo glanced over at the Chinook and saw that it was dangerously close to the downslope. Apparently, the pilot saw it as well, because the chute drifted east again a full fifty meters or so as the chopper moved east. Then the guy in the open door cranked the line back into the Chinook fifty meters, so the chute once again was directly below, while the Chinook was fifty meters further east. Then, ever so slowly, the pilot lifted the bird, bringing the chute closer and closer to Dybo. Dybo reached out, but missed it as a gust of wind moved it a meter down. The chute eased up again, and this time, Dybo grabbed it and hauled it to the platform. They had to be very careful, since any major shifting of the Chinook while the line was still connected could spell disaster.

Dybo passed the line around a pulley and up to Chrietzberg, who wrapped two turns around an electric winch. As Chrietzberg ran the winch, he let the wind in the chute pull the line out and away. The end of the pilot line attached to the lightweight line left the chopper door, and the heavier line dropped into a catenary as it traveled the distance. As the pilot line approached the platform, the Chinook began a very slow descent to give a more vertical aspect to the pilot line. The pilot line passed over the pulley and up and around the winch. Chrietzberg let it continue for another twenty meters, and then stopped the winch and cinched the cable, winding it around a cleat. He pulled the bitter end from the sky behind him and cut loose the lightweight line. The line disappeared immediately, flung away by the 60-knot wind.

While Dybo waited at the end of the platform with a sharp knife poised to slice through the Kevlar line should anything go wrong, Chrietzberg hauled the end of the pilot line up to the terminus, around the pulley, and over to the winch, where he took several wraps, and tied off the bitter end.

"Hey, AJ. I need you to take the pilot line off the winch and cleat. I'm ready up here."

"Got it." Dybo sheathed his knife and climbed to the winch. He took the line off the cleat and then slipped it off the winch. It pulled back with a snap. "Take up the slack, Critz."

"I'm on it."

As they accomplished this, the Chinook bay attendant, who had been monitoring their conversation, dropped the outer face of the reel, and the remaining pilot line tumbled into the wind, and was immediately carried away from the helicopter. Simultaneously, the Chinook dropped down and away, and headed back to Baker.

Hauling in the pilot line was not particularly difficult. They let the accumulating line trail in the wind. The only problem was that the task was slow. Their haul rate was about 2.5 meters per second, so they were at it for an hour, before the skytower cable finally hove into view, bringing with it a second pilot line. Attaching the skytower took an hour. The attachment held the weight of the entire skytower bundle—all eighty kilometers. In effect, when Dybo and Chrietzberg were done, it had become an integral part of the track itself. With that accomplished, they commenced hauling in the next pilot line with its deflector payload. An hour later, they began attaching the deflector to the terminus. Like the skytower cable, it became an integral part of the track.

The deflector arrived with a pilot line passed through it and leading back down to Baker. Once the deflector was securely attached, Baker commenced pulling the pilot line through the deflector. They did this at a much higher rate than Dybo and Chrietzberg were able to on the skyport. In less than a half-hour, the lift cable was in place. This signaled the commencement of the ascent back into space.

While Klaus' linear drivers began to increase their load, running the ribbon ever faster, Dybo and Chrietzberg commenced disassembling the platform they had built only three days earlier. Since they didn't want to drop the pieces, even though the chances of their striking anyone or anything were remote, the task went slowly. They removed a piece, hauled it into the skyport, removed another, and so on, until they finally reached the last few pieces directly beneath the terminus. Chrietzberg grabbed a stanchion, unhooked his safety line from the remaining platform, and attached it to a ring extending from the bottom of the skyport. Then Dybo grabbed the same stanchion with his left hand, and unhooked his safety line with his right.

At that moment, the entire skyport gave a mighty shake as a sudden gust of wind slammed into it. Dybo missed the ring with his safety line carabiner as his body swung away from the platform. "Help me, Critz!" Dybo shouted as he grabbed for a stanchion with his free hand, missing. The skyport shook again sharply as another gust passed. Dybo strained with every bit of strength he had to extend his hand to Chrietzberg, who reached out with both his hands, trying to catch Dybo.

"I'm losing my grip, Critz! I can't hold on!"

Another shake, and Dybo lost his grip. "Ohmygod…Critz…I'm falling…"

Almost instantly, Dybo lost his feeling of weight. He turned to see the skyport disappear above him off to the west as he plummeted

Earthward. Oddly enough, he did not have a sense of falling so much as a feeling of floating. He thought of the holocasts he had seen of skydivers doing all kinds of stunts in the air. *I got nothing better to do*, he thought as he twisted and turned in the air, trying to find a stable position. As he fell, the air around him thickened causing him to feel the increasing pressure of the air against his suit as it rushed up to meet him. The wind had carried him to the east, and the storm was directly below him, spread out in all directions. The roiling clouds did not form a smooth surface. Near the leading edge of the storm and again near the leading edge of the eye, massive thunderheads pushed their way to his present height, dwarfing him as he plummeted downward, arms and legs spread-eagled into the wind. He wrapped his hand around the manual chute release in case the automatics didn't work. One moment he was mostly above the storm with thunderheads for company, the next, he had passed the edge of the eye and into the relatively clear center. His horizon became limited by walls of cloud in all directions, clouds that got darker and darker as he fell. The sky immediately above him remained blue, but the only thing below was black, wave-tossed water. He had no reference to determine wave height, but they looked big, and they were rushing up to meet him.

With a pop and a jerk, his chute opened and blossomed into a shaped canopy over his head. Weight returned, and with it a very real sense of hanging from a harness. And then he saw how close the waves were, and then his booted feet hit the water, and his chute automatically detached. He sank into the water so that it covered his helmet, and all he saw was green-blue water everywhere. He began to feel a bit of panic rising from his inner seat of consciousness, imagining himself continuing down to the seafloor, another five kilometers below. But shortly, his descent stopped, and he popped up out of the water to his chest, and remained there, half in and half out of the water. Struggling was useless, and there was no way he was going to remove his suit. He tilted himself back, and found that he could practically lie on the surface. He lay there, glad to be alive, not entirely sure how he got there, wondering what was next.

"AJ, it's Margo. We've got your position pinpointed. We're on our way and will reach you within the hour." There was a pause that seemed to last forever. "Don't go anywhere!"

CHAPTER THIRTY-ONE

BAKER ISLAND—CONTROL CENTER

"Mayday! Mayday! Boss…we just lost AJ!" Chrietzberg's words tumbled over each other.

"We got you, Critz. What happened?" Alex answered immediately. "Slow down and tell me what happened."

"We wuz switching safety lines from the platform to the skyport, and a big gust hit us. AJ didn't have a chance. I couldn't hold him. My God, Boss…I couldn't hold him!"

"We got his beacon. He's headed for the water southeast of Baker, Critz. Margo's people are already on their way to get him."

"No shit! He okay?"

"We don't know that, but his beacon is strong. We'll keep you informed." The wind commenced howling at that point, nearly drowning out the conversation. "We got to get you back to height, and we got to fight a storm down here. Hang tight. We'll get him."

The Chinook landed on the apron just outside Baker Compound in eighty-knot gusts. The crew tied it down while the pilot, Chad

Mickey, hustled inside. He doffed his ever-present Stetson, using it to brush the rain off his lanky 178-centimeter frame. Even so, his jeans and boots remained soaked.

"What happened up there, Chad?" Alex asked him.

Mickey's hazel eyes twinkled with excitement. "Don't know, Sir. When I peeled away, they were doing just fine. Those guys got some balls, I got to tell you."

"You didn't do so bad yourself," Alex told him with a warm smile. "That was some flying."

"Like rockin' a baby. But that's a great bird—steady as an eagle."

"Hang around for a couple more days," Alex told him. "We're going to need your services one more time when the track reaches seventy-two klicks. We'll need you to carry the double deflector from the barge to the socket. I'll give you an hour or so notice. Until then, just take it easy."

"Can I hang here?" the pilot asked.

"Sure. Just stay out of the way."

"I'll keep him busy for a while," Lori said as she swept into the Control Center with Dex in tow. "I'm Lori from the Fox Syndicate. This is my cameraman, Dex Lao. You're the Chinook pilot, right?" Lori examined Mickey from head to foot, and then stepped toward him with moistened lips.

"I…I'm Chad Mickey, and…I'm…I…I fly that ugly bird."

"You okay with an on-camera interview, Cowboy?"

"Sure, Ma'am…I mean Miss Lori."

"I'm Lori, Chad." She smiled warmly at him, and glanced at Dex, signaling that he commence recording. "Tell me about your bravery out there."

"Not me, Lori. I just flew the Chinook. Those guys up there, those two Mohawks, they are the heroes. I never saw anything like it."

Alex watched with amusement as Lori quickly wrapped the big pilot around her little finger. It never failed. This lady used sex as naturally as she breathed, without offending. She could come on to a man and to his date, and they both would love her for it.

Margo approached him from behind and placed her hands on his shoulders. "Stop leering," she whispered into his ear. She put an arm around his shoulders and walked him toward the holodisplay covering the wall. "She really gets to you, doesn't she?"

"Me and everyone else," Alex answered with a grin.

"I thought I was your flame," Margo said wistfully. "No way I can compete with Lori, though. Hell, she gets to me too!" She laughed lightly and stepped away from Alex's side. "Looks like we're all working," she said with another amused glance toward Lori doing her interview.

The storm was not having much effect on the ascension. The ocean-going tugs at the Eastern and Western Complexes were pushing the apexes outward, but the main action was split between the barges in Meyerton Harbor at Baker and Millersville Harbor at Jarvis, and the Baker Socket. At Jarvis, the action was minimal. The Skycrane had faked the cable bundle down in a way that allowed it to be pulled off the pile and over the large cushioned roller without tangling, and then across the A-frames to the socket. The Skycrane and support people were on standby just in case, but thus far, it had gone like clockwork. The Baker operation was more complicated, because the skytower cable was being fed from a giant reel in one barge across the A-frames, while the suspensor and lift cables were being fed from the open hold of the other barge across the second set of A-frames. They were married at the socket exactly as they were during the original Ascension. Furthermore, to lighten the Chinook's load while hauling the deflector to the terminus, the first eighteen kilometers of lift cable were replaced by pilot line. Once the deflector was attached to the terminus, they had to pull the lift cable up around the deflector and back down—all eighteen kilometers. They married the two ends of the lift cable at the double deflector. When the skyport reached seventy-two kilometers, the Chinook would carry the double deflector to the socket where the crew would marry it to the skytower.

The rail bootstrapped itself skyward with a deliberate pace, mimicking the original Ascension as the storm finally moved further west and dissipated. As the two downslopes lifted out of the water, exposing for a second time the tensioner attachment points, the ten payout reels once again fed cable in their delicate dance of balance-counter balance, keeping the downslopes and rail perfectly plumb. As the rail rose, the three pairs of tensioners along the horizontal length extended themselves again from the abyssal plain five kilometers below the surface, maintaining the rail's verticality. A more relaxed general demeanor filtered throughout the 2,500-kilometer leviathan, since the process was now familiar territory. As before, the process was primarily monitored by computer, since reaction time to any unforeseen event was time-critical. By the time Alex would have recognized a problem and taken corrective action, Slingshot would have torn itself apart.

    In the Baker Operations Center, Alex kept his finger on the pulse. Every five hundred meters of elevation, he stopped the Ascension for the installation of the rings and their associated tubes and cables. Actually, the computer controller stopped the ascension, but Alex monitored it closely. As with the original Ascension, during the half-hour or so that this operation required at Baker Socket, the ocean-going tugs topped off their fuel, and people throughout the project inspected and reported to Alex. As always, Alex relied on his computers, but depended on his people. When the shit hit the fan, as it had twice already, people solved the problem, not the computers. As a fully integrated person in the complex modern world, Alex completely understood the ubiquitous role computers played in every aspect of life, but he never lost sight of the human element and the importance of each human being in the overall process.

    Ascension speed was determined by the rate at which the ribbon could be accelerated and the frequency and duration of the tasks that interrupted the rise. With seventeen daily half-hour holds to install the spacers, the maximum rise rate was nine kilometers a day. The process was inexorable, requiring the full attention of a large number of people twenty-fours a day. Alex and Margo took to spelling each other, and on Jarvis, Cody Haydon joined Klaus to give him some genuine off-time to catch up on his sleep. Since the outward appearance of the operation seemed not to change much, Lori had nothing to report, and so spent a lot of time getting to know Chad better. Alex noticed that whenever he saw Lori and Chad, Noe seemed to be there as well. It conjured distracting images that he thrust aside. *Can't go there…too much going on here!*

BAKER ISLAND—UNDERWAY ON *SKIMMER THREE*

Pearl Wells had been eagerly following the action ever since the explosion at *Amelia Earhart Skyport*. Virtually every moment she was not in the water or accomplishing some other necessary task, she was glued to her Link. She cheered when they captured that Watson bastard, she cried tears of happiness when the stranded high-iron guys were rescued, and she whooped with unrestrained joy when the Chinook hooked up with *Earhart Skyport* nine kilometers above Baker. When Dybo fell off the platform beneath *Earhart*, Pearl was getting *Skimmer Three* underway even before Margo called her.

"Take at least two guys with you," Margo said. "And, for heaven's sake, be careful. It's nasty out there. I don't want to mount another search and rescue because you got into a jam."

"Pu-leeze!" Pearl responded with good humor. "I don't get into jams. You know that, Girl!"

"Just the same," Margo said, laughing, "take it easy. AJ's good for several hours out there—if he's alive. His comms're out, so we don't know. The only thing we have is his beacon. How's that signal on *Skimmer Three*?"

"It's fine, Margo. Now, are you going to let me do my job, or are you going to drive this sucker by remote control?"

"No problem, Pearl…keep me in the loop, okay?"

"Roger that."

The storm center had passed, and rain pelted down with almost stinging force, driven by winds still gusting to fifty knots. Even though it was well into the morning, the sky remained twilight gray.

On her way to the skimmer, Pearl had grabbed Domingo Solak and Abel Kilker. Just as she was pulling away from the dock, Emmett Bihm leapt across the widening gap.

"Glad you could make it, Bimmy."

"You the Man, Pearl!" Everybody laughed, and then Bihm said, "This could be tough, so…Pearl, you're the skipper—you got the boat ops. I got the water ops. Everybody clear on that?"

"Good thing you got here, Bimmy," Pearl said with mock seriousness, "otherwise I'd have to do both. Don't know if I cudda handled all that responsibility." Then she turned her attention to navigating out of the artificial harbor into the turbulent waves beyond the breakwater. The skimmer was pitching and rolling like a cork in a bathtub with a splashing rugrat. There was no way to get up on the cushion. Pearl had to push her way through the waves with brute force. She turned to port as a big wave lifted the bow until the vessel almost stood on its stern, and then pounded back into the water with a force that nearly knocked the air from her lungs. She twisted the helm to starboard to take the next wave at an angle. That worked much better as she rode up the wave face, over the crest, and slid down the backside. The divers whooped and hollered in delight as they grabbed handholds and rode out the water-borne rollercoaster.

"Hey, Critz—it's Pearl on *Skimmer Three*. Thought you'd like to know we're on our way to AJ."

Chrietzberg responded immediately, obviously having been monitoring the links. "Hey, back. Thanks for calling, Pearl. Keep me in the loop, okay? Keep the Link open, so I can lurk. I'll stay out of your hair."

"Sure thing, buddy. We gonna get that boy." She left the Link open as he asked. "Critz is with us, guys," she said to the others, pointing to the Link. She was concerned that if they found the worst, the guys should watch their comments. She got thumbs-up from all three, and immediately turned her attention back to the next wave. As she crested the peak, she shouted, "This is a bunch of crap! We gotta get into Baker's lee, or we're never gonna make any headway." She turned starboard as she crossed the crest, and gunned the skimmer sideways down the wave and along the trough. As she picked up the next wave, she turned to port, into the wave, and as soon as it crested, she spun to starboard and gunned the skimmer again, gaining several hundred meters before the next wave picked her up. The waves were refracting around Baker, hitting her from the south, as she ran to the east along their backsides and troughs, being careful not to get caught on Baker's northwestern reef. A half-hour's hard work got the skimmer around Baker's northwest corner and into the lee. The water settled down to a more navigable surface, and Pearl set course for the beacon flashing on her console some thirty kilometers off the southeastern corner of Baker.

"That was some boat driving, Girl," Bihm told her as things settled down. "You done that once or twice."

"I guess," Pearl answered, inwardly pleased that she could show these guys a thing or two about boats. She powered the skimmer up onto the cushion and cranked her up to seventy-five knots. The water was too rough for higher speeds. *Hang on, AJ. We're coming. You be alive, hear? You be alive!*

Thirteen minutes later, Pearl throttled back and let *Skimmer Three* settle into the water. Then she eased forward at a bare five knots. "Get your eyes open, guys. He's just off the starboard bow, a hundred meters or so."

"I got him!" Solak shouted, pointing through the window. "Sonofabitch is waving at us!"

"I got him too," Kilker said, stepping into the stern well, ponytail whipping in the wind.

Pearl brought the skimmer as close as she dared. "Can't get any closer, guys, without some help in the water."

"I agree," Bihm said. "Skimmer's pitching too much to haul him out. We need someone in the water. That's you, Abe. You're the strongest," he said to Kilker. "Skins, fins, mask and snorkel…and safety harness."

During the short transit, both divers had donned their skins, so Kilker was in the water almost immediately, trailing a line attached to his harness, and another for Dybo. He wrapped the line under Dybo's arms leading away from his back, so that Bihm and Solak were able to pull him to the diver platform along the stern. With Dybo facing away from the skimmer and Kilker assisting from the water, Bihm and Solak hoisted him onto the platform, and then up and into the stern well. Kilker hoisted himself aboard without assistance, and they carried a struggling Dybo inside the skimmer.

Dybo reached up and grabbed his helmet, giving it a half-turn to the left. With a soft hiss, the helmet broke its seal, and he removed it. "Goddamn sonofabitch…I made it!" he yelled at the top of his voice. "I'm alive, guys…I made it!"

"Hey AJ, it's Critz. You made it, you lucky bastard! You made it!"

"You got that right, cousin…you got that fucking right!"

Pearl smiled as two tears trickled down her black cheeks. *You the Man, Girl,* she whispered quietly to herself, *you the Man!*

BAKER ISLAND—CONTROL CENTER

During the following nine days Lori interviewed Dybo twice, once by himself and once with Chrietzberg participating by link. The entire world had followed her tense reports from the scene of the disaster, and the world celebrated the Mohawk hero who fell from the sky and lived to tell the tale. Stories about the Kahnawak Mohawks appeared in holocasts and publications the world over. Special interest groups formed across the Web, propagating the interest in everything Mohawk. Knowing how to say something in the Mohawk language became an instant rage. *Kwe* (hello), *Onhka ni:se* (how are you), and *O:nen* (good bye) replaced local greetings in virtually every country in the civilized world.

While biding his time in the ascending *Earhart Skyport*, Chrietzberg was besieged with interview requests and received hundreds of marriage proposals from around the world. Dybo fielded so many appearance requests, and ducked so many marriage proposals that he asked Alex for a week off, just to sort things out. He promised, however, to be back before Chrietzberg returned to the surface.

Pearl also found herself at the center of a small media storm. Once the media discovered her role in the rescue, they deluged her with interview requests as well. And that trickled down to the other three who accompanied her.

Alex finally had to issue a general announcement that there would be no more contact with outside media, except through Bruce and Lori, who would handle all requests, and hold the circus at bay while the Slingshot team completed their task of reestablishing the rail and *Earhart Skyport*.

On the evening of the ninth day, beneath a darkening, nearly cloudless azure dome filled with skreeghing terns, boobies, and lesser frigatebirds, Lori turned to face Dex Lao's holocams. Behind her, as the entire world watched, Rodney Chrietzberg stepped out of the capsule that had just arrived from *Amelia Earhart Skyport*, and gave his cousin Ajay Dybo a high-five.

CHAPTER THIRTY-TWO

SEATTLE—DOWNTOWN

Neil Lansing, a.k.a. Lars Watson, stepped off the SeaTac people-mover at the main terminal and rode the escalator to street level. He wore a suit that really was gray and carried a black briefcase. His dark hair was cut in the current fashion, not too long and not too short, tapered on the sides and back. He sported a dark full beard, neatly trimmed. There was virtually nothing to distinguish him from the dozens of other businessmen arriving in Seattle that morning.

The escalator deposited him on the terminal main floor, where he exited the secure perimeter and looked around expectantly. Near the door, he spotted a twenty-something man holding aloft a sign that read "Lakeview." That was his man. Watson strode toward him and said, "You have something for me?"

Without a word, the man handed Watson a manila envelope, set his sign in the corner, and left through the sliding door to the street. Watson followed him through the doors and looked for something indicating public transportation to the city. It was obvious and immediate, and five minutes later, Watson relaxed in the back row of a comfortable bus for the half-hour ride into the city. He looked out the window curiously as the bus navigated a complex series of ramps and tunnels that led to the Interstate-5 corridor and onto a dedicated bus lane. The sky was peppered with puffs of white clouds that seemed to be moving toward Mt. Rainier, dominating the eastern horizon. To the west across Puget Sound, storm clouds appeared to be pushing up over the snow-capped Olympic Mountains. A storm was on its way and probably would hit the city by nightfall.

The Seattle skyline with an aging Space Needle dominated the northern view, and as the bus turned off the freeway onto the surface streets of the city, the Space Needle was swallowed up by the plethora of shapes and sizes pushing into the sky around them. The bus stopped at the downtown airline terminal to disgorge its passengers. Watson strolled out to the street and looked around. The terminal was a couple of blocks north of the distinctive Smith Tower. Several boutique hotels lined the boulevard, and he entered the lobby of the first one he encountered, the Sinclair Arms, a five-story 100-room hotel that catered to downtown business visitors.

Watson still had about four hours before he would purchase a cup of coffee in the Smith Tower lobby. His old undergraduate *alma mater*, Evergreen State College, still a veritable hotbed of progressive politics, was located about 120 kilometers to the south in Washington's State Capital, Olympia—too far to visit. University of Washington, his graduate *alma mater*, on the other hand, was just eight kilometers north on Interstate-5. Watson called a cab and told the driver to take him to the UDub Information Center. About twenty-five minutes later, Watson paid the driver and stepped onto Campus Parkway. He walked across Fifteenth and entered the campus. Not a lot had changed. He walked east past Meany Theatre and into Red Square. He angled left and strolled along Pierce Lane through the Quad. The pale red brick buildings, appearing mottled gray to Watson's eyes, were well over a hundred years old. Their tall, narrow windows looked out over a tree-filled park crisscrossed with pathways. He well remembered walking these pathways as a student, sitting in many of the classrooms overlooking the Quad, filled with a sense of youthful purpose. He turned right to walk past Thompson Hall and continued down to the Husky Union Building, the HUB, with its open green where he used to exhort crowds of eager students to "burn, baby, burn" the contraptions of modern life that were destroying the planet. As he walked along the front of the Student Union Building, dozens of students occupied tables scattered across the pavement, sipping drinks, eating late lunches, studying, arguing. A rush of nostalgia filled Watson's being as he stepped back into the shadow of a tree to watch. A young radical with long unkempt hair and ragged shirt and jeans waved his hands in the air to emphasize his passion of the moment—something to do with military recruitment. Watson resisted the urge to step to the platform himself, to tell the assembled crowd about Slingshot and how it threatened the world—and what he was doing about it. His heart ached as he watched the pretty girls with their short skirts, crossing and uncrossing their legs, badly needing some guidance from a man of passion and experience…but he had a job to do. He checked the time. It was time to leave this small paradise and return downtown. Watson worked his way north to 45th and 15th, where he hailed a cab.

The cab dropped him off at Second Avenue across from the Smith Tower at 4:25, just nine minutes from his designated "appointment." He crossed at the light and sauntered into the ornate lobby just after 4:30. At exactly 4:34, Watson stepped up to the *Seattle's Only Java* cart and said, "I'll have a java with double cane." The cart operator looked at him with interest, reached under his counter, and handed Watson a paper cup just under twenty centimeters tall. It looked exactly like cups being carried through the lobby by other customers. It was warm to the touch, completely innocuous. Watson left five dollars on the counter.

Don't drink it, he had been admonished. So he didn't.

Back in his room at the Sinclair Arms, Watson poured the coffee into his sink and retrieved the two vials of liquid. Holding them up to the light, he examined them carefully. They were completely full with no air gap, and were sealed at both ends. Out of curiosity, he filled a water glass from the sink and dropped a vial into the glass. Sure enough, it disappeared completely. There was not a hint that the glass contained anything but plain water. Watson opened the small refrigerator by his desk and took a bottle of spring water. He opened it, poured half the water down the drain, and dropped both vials into the bottle. Then he topped it off from the faucet and replaced the cap. It was indistinguishable from a bottle of plain water. Satisfied, Watson put the bottle back into the fridge and went down to the hotel lobby.

It was still early, so Watson decided to walk to Pike Place Market, about a mile north on Second. The produce and fish vendors would be gone by now, but he had always loved to wander through the maze of passages, stopping at a used book store, or examining the wares of a candle maker, old coin shop, or purveyor of exotic spices and teas. He ended up sipping a strong, orange-flavored, spice tea in a small parlor overlooking Elliot Bay, reading a book he had found in a dusty antique book den. The book told the story of the founding of Greenpeace in 1971 by the ecologist Patrick Moore, and how he left the organization in the 1980s to pursue his belief that nuclear power was the way to go. Watson viewed him as a seriously flawed hero who ultimately abandoned everything that mattered. In his mind, Watson saw himself as the polar opposite. As the evening sky over Elliot Bay darkened into night, assisted by the looming storm clouds that had worked their way across the Olympic peninsula, Watson sensed that the tea parlor keeper wanted to go home. He purchased a bag of the heady tea leaves and a single cup infuser to take with him to Baker Island, said good night to the shopkeeper, and found his way through the warren of passageways back to Second Avenue, trailing a lingering aroma of exotic spices and incense in his wake.

The next morning at 9:30, Watson entered one of the ornate express elevators in the Smith Tower lobby. Curiously, the elevator was manned by a human operator, a throwback to a century-and-a-half ago, and probably the only such operator anywhere in America. The operator was an attractive nineteen-year-old college girl from UDub. She wore a vintage, lace-trimmed, light-green city gown from the late nineteenth century featuring a low décolletage, short puffy sleeves, and over-the-elbow white gloves. Although Watson remained unaware of her outfit's light-green color, he very much got the décolletage. She was pert and sassy, and flirted with him all the way to the thirty-fifth floor. The elevator opened to an embossed copper mural of Slingshot, vertically exaggerated to show its details. Watson went cold when he saw the mural, pasted a smile on his face, and turned to greet the receptionist.

An hour later, Watson had his reservations and tickets, and knew his schedule at Baker and *Amelia Earhart Skyport*. *LLI* was efficient, he had to grant them that, but his skin still crawled as he entered the express elevator. The operator this time was a frumpy, pimply lad of eighteen who seemed completely out of place in his dapper plaid, high-buttoned suit and vintage rounded-crown derby. Watson stood at the rear of the car, ignoring the lad. On his way through the lobby, he purchased a real cup of coffee and sat down at a small cast-iron-and-glass table to enjoy it, letting its warm aroma soothe away the tension of the past hour. Two men passed his table, apparently heading for the elevator bank he had just used. He examined them curiously. They were dressed in normal business attire, in variants of gray as Watson perceived them, but they both carried themselves with a confidence that suggested they didn't spend all their time behind desks. As they passed Watson's table, they looked right through him. Watson didn't mind, though, and finished his coffee in a leisurely manner before tossing his empty cup in the recycle bin and strolling out into the rain.

SEATTLE—SMITH TOWER

Quinton Radler strode into the *Environment, Inc.* meeting room from his private door on the fourteenth floor of the Smith Tower as he had on so many other occasions during the past two years. Dane Curvin, his personnel head, Katelynn Leete, head of manufacturing sabotage, and Darius Gotch, in charge of site sabotage, were waiting for him at the conference table. He had assembled them to brief them on the status of their latest attempt at bringing down Slingshot.

"Watson will be flying to Baker tomorrow morning," he said. "He's carrying the gel-pack, and I really believe he will pull it off this time." He looked at Gotch. "Darius…"

"It's simple, really," Gotch said. "He is set to do an inspection that includes the outside. He should have plenty of opportunity to mix and place the gel. Since the suits all carry chutes now, he's planning to jump right after placing it."

"And then?" Leete asked. "Do we pick him up or what?"

"That's the cool part of this," Gotch answered. "He has no idea who we are, none whatsoever. Soooo…we don't do anything at all.

He makes it back down, they pick him up and nail his ass." Gotch clasped his hands behind his head, leaned back, and smiled broadly.

What a jerk, Radler thought as he looked at the others. I will be so glad to get the fuck outta here after the crash…go do something constructive instead. I'm sick of this shit. He looked up as the door from the front office suddenly opened. What the fuck…

"What is this?" he demanded, his voice filled with anger, as two men entered the room. They were dressed in business attire, but they were taller by several inches than anyone in the room, and larger across the chest. They placed themselves at two of the doors, and the man by the office door presented a leather wallet.

"I'm Deputy Federal Marshal James Franks, and this," he indicated the other man, "is my partner Deputy Marshal Ronald Botts. I have federal warrants for Quinton Radler, Darius Gotch, Dane Curvin, and Katelynn Leete." He held up his hand. "Remain seated, and don't move! Fifteen of Seattle's finest have locked down this entire floor."

Radler looked around the table in exaggerated shock. "What do you mean, federal warrants?" he said, rage creeping into his voice. "This is an environmental consulting firm. How dare you burst in here waving your badges around like a couple of cowboys."

"Quinton Radler," Franks said back, calmly ignoring the outburst, "you're under arrest for aggravated terrorism, and conspiracy to commit murder." Then he turned to Gotch. "Darius Gotch, you're under arrest for aggravated terrorism and conspiracy to commit murder." He turned to Curvin. "Dane Curvin, you're under arrest for espionage, aggravated terrorism, and conspiracy to commit murder." And finally, he turned to Leete. "Katelynn Leete, you're under arrest for espionage, aggravated terrorism, and conspiracy to commit murder."

Botts walked around the table, securely cuffing their hands behind each of them. "You have the right to remain silent," he intoned. "You have a right to an attorney…"

BAKER ISLAND

Alex stood on the berm overlooking Meyerton Harbor to the west and *Baker Skytower* to the east just beyond the Operations Center. His eyes followed the berm to the north past the Dive center, and around to Margo's bungalow with its spectacular view of the reef. The landing strip to the south occupied pretty much the same spot as it had for over a hundred years, although until Slingshot arrived, it was virtually never used. The buildings clustered around the socket had all been rebuilt. And the skytower stretched into the blue sky, disappearing before the eye really had a chance to bring it into focus. Alex let his mind wander over the many events of the past two years. He found himself coming back again and again to the Hyperchess games he played with Margo during the earlier phases of the project, their times together on the beach, diving on the submerged rail as they investigated the early sabotage incidents. He missed those times, but with the problems that Watson had brought into his life, and with the increasingly complex character of the project itself, Alex found little time left over to consolidate his gains with Margo. He knew she was spending more and more of her free time with Klaus, and he really had no problem with that. Klaus could not have been more different than himself, and Margo was more than competent to decide her own future. He grinned as he recalled Margo's tryst with Lori, and wondered how that went down with Klaus. They had never discussed it. Lori—now there was a woman to get to know better. She had telegraphed her interest to him so many times that they were almost like an old married couple, and yet he had never seen her with her clothes off. Poor Mary Martain—what a tragic way to go, and what a way to lose a lover. Alex could not imagine how he would deal with having an intimate partner literally ripped from his arms in death. But Lori was as tough as they come, no matter the outward appearance. She bounced back with Noe and Chad—now that had to be interesting...

"Boss," Bruce Yoon said, as he approached Alex from the launch facility at the socket, "I've been trying to reach you, but your Link seems to be off."

Alex switched his Link back on and said, "It's on now, Bruce. Shall we go?" He put his arm around the young man's shoulders, and they strolled down the backside of the berm to the small crowd waiting at the launch ramp. "I guess it's time to get this show on the road."

The crowd consisted of virtually all the main players during the last two years. The dive teams were present, except for the duty section, and they were participating by Link. The pilots were there, as were the coxswains, the crane drivers, in fact, virtually everyone not specifically on duty who could make it from wherever they normally were to Baker—they had joined the crowd. Lori was playing the group, talking on camera with as many individuals as she could while waiting for Alex to show. As Alex and Bruce approached the crowd, Apryl Searson broke away and ran to Bruce, throwing her arms around his neck in a display of unabashed passion. She stopped momentarily to look at Alex and say, "Hi, Boss. How're the ribs?" But before he could answer, she put another liplock on Bruce.

Neil Lansing, a.k.a. Lars Watson, was also part of the group. He was designated to be on the first capsule, so he could complete the first formal FHWA inspection before the official dedication ceremony on *Amelia Earhart Skyport's* main deck. Because Lansing would conduct part of his inspection outside the skyport, Alex designated Ajay Dybo as his partner while they were outside. Following the inspection and dedication, Alex planned to run a capsule rail abort to atmosphere test. If that went as planned, *LLI* would spend the next few hours giving free rail rides to Noonan for those project members who wanted to make the trip.

Alex watched as Lansing and Dybo climbed into the waiting capsule, followed by Lori and Dex, Margo and Klaus, and finally Bruce. Then he turned to the group and waved to them as they cheered enthusiastically. He quieted them down and said, "Sorry you can't all come up with us on this first trip, but the dedication will be broadcast, so you can see it right here, and all of you who wish to will come up in the next few hours—I promise." He closed the gull-wings, and two minutes later, the capsule disappeared into the nearly cloudless sky.

Five minutes later, as the eight passengers disembarked eighty kilometers above Baker, Alex signaled Dybo to get Lansing suited up for the external inspection.

AMELIA EARHART SKYPORT

Dybo walked Lansing through the lock and sealed it behind them. Lansing was carrying the water bottle that he seemed to have with him virtually all the time. He laid out a suit for the inspector and told him to remove his shoes and jacket. As Lansing removed his shoes, Dybo could not help but notice that his socks had identical patterns, but they were entirely different colors, one green and one red. It was obvious that Lansing was unaware of this fact. Dybo said nothing, and turned to lay out his own suit. Right after Lansing slipped his legs into the suit, but before he slipped his arms into the top, he opened his water bottle and took a swig. Although Dybo wasn't watching him very closely, he could have sworn that the guy took something from the bottle and put it into his leg pouch. Then he noticed that the inspector was adjusting his check-off list in his leg pouch and thought nothing more about it. Even so, the mismatched socks continued to bother him.

They finished suiting up, and Dybo attached Lansing's helmet and then his own. When he was certain they were sealed, he closed the capsule hatch and remotely moved the capsule out of the way. He evacuated the chamber with the pumps while Lansing did his chamber inspection. Once the chamber was evacuated, Dybo cracked the capsule hatch to let out any residual air, and then opened it. Lansing gingerly moved along the track, checking off his list as he went. When Lansing reached the launch bay with the rail centered in the bottom, Dybo watched him reach into his leg pouch and remove something that looked like a collapsible cup. *What the hell!* Lansing then reached into his leg pouch again and removed what appeared to be a glass vial.

Suddenly the picture began to gel in Dybo's mind. He distinctly remembered sitting in the Baker canteen sharing a beer or two with his fellow high-iron workers. One guy in the group was different. He didn't really belong, and it turned out he was an inspector. He was a strange guy who sometimes wore incredibly mismatched clothing, with totally contrasting colors. It actually became a bit of a joke with his Mohawk friends. Pete LaFleur told him about the mismatched socks, and said that he had onloaded the kick thrusters with that guy, and that he had disappeared right after the explosion. Dybo thought his name was Bill or Bob Wein…something or other. Then recognition dawned. *Sonofabitch! This Lansing guy IS Weingard. He's not an inspector, he's a saboteur!*

Although Dybo had no idea what Lansing was up to, he was convinced it was to no good. "Hey, you! Lansing! What the hell are you doing?" he shouted over the circuit. Lansing completely ignored him and reached into his leg pouch a second time. Dybo made an instant decision. He lunged at Lansing, grabbing him by his backpack, twisting him around. Lansing kicked toward his groin, but as Lansing was off-balance, the kick was ineffective. Lansing pulled a long screwdriver out of his leg pouch and threw himself at Dybo. Dybo warded off Lansing's blow and grabbed his utility knife from his leg pouch as Lansing twisted away. Dybo lunged at Lansing's back, slicing the chute case mounted above his air cylinders. Then he grabbed him by his cylinders and twisted him away from the rail. As they struggled, he lost the utility knife, and Lansing dropped a glass vial onto the blast shield beneath the rail. Surprisingly, it didn't break. Dybo attempted to pry a second vial from Lansing's gloved hand. As he did so, he felt Lansing's screwdriver pierce his suit arm just below his right shoulder, and penetrate his bicep. He let go of Lansing's hand containing the second vial, reached up with his left hand, grabbed the screwdriver handle, and yanked it out. He felt his suit seal the rip as he tossed the bloody instrument out the launch-bay opening where it disappeared from sight. Lansing turned in what appeared to be an attempt to retrieve the fallen vial. Dybo swept his feet out from under him with a round-house kick.

Lansing landed astride the rail with a look of total astonishment on his face. A reinforcing band was just ahead of him. He reached down to the blast shield, grabbed the fallen vial, and began fumbling with the two vials. As Dybo watched Lansing's contorted face, he suddenly realized that it really was the same guy. *You sonofabitch...all you did was dye your hair and grow a beard.* Apparently frustrated by his attempts to open the vials, Lansing placed both in his left hand and tried to snap off their tops against the edge of the reinforcing band. Realizing this was the critical moment, Dybo threw himself against Lansing. Lansing dropped the still-sealed vials onto the blast shield and fell back against the rail, gripping Dybo with a death grip. The force of Dybo's impact slid them along the rail. Dybo grabbed the rail with both arms, as he felt himself sliding off, forcing himself to ignore the stabbing pain in his upper right arm. He slid down, angling from the rail so that only his clenched fingers atop the rail kept him from plunging to Earth. Lansing slid off the rail on the other side, clawing for Dybo's arms as he did so. For a moment, Dybo's clenched fingers were holding both their weights as Lansing tried to crawl up Dybo's body with his legs. Dybo twisted and turned, willing his gloved fingers to remain clenched, and finally shook off Lansing's legs. That did it. Lansing lost his tenuous grip on the rail, and in an instant, Neil Lansing, a.k.a. Lars Watson, was tumbling through the rarified atmosphere toward the ocean eighty kilometers below.

Dybo slowly inched his way back along the rail until he could place his feet on the blast shield. Carefully he moved away from the opening, and then leaned down to retrieve the vials. "Boss," he said into the circuit, "I'm in trouble. I'm going to need some help out here." And he collapsed on the blast shield.

※

While Lansing and Dybo were outside, Alex moved off to a far window by himself and gazed out into the star-studded sky above him. Mabel had already discussed with him what the next project would be. She asked him if he would like to build the first tethered sling ever constructed, out past geosynchronous orbit, positioned to catch capsules from Slingshot, and send them on their way to the Moon, Mars, or the Asteroid Belt beyond. He had agreed before she completed her sentence. He gazed into the distance and shuddered as a wave of emotion swept through his body. It was a project that dwarfed Slingshot, and he would make it happen. He turned to see Margo and Klaus across the main deck in a loving embrace that broadcast their intentions. Clearly, Margo had finally chosen. She opened her eyes and looked back at Alex. She mouthed the words, *I'm sorry*, and then she closed her eyes and kissed Klaus again.

Alex felt a touch on his shoulder. He turned to find Lori standing behind him. She had seen his silent exchange with Margo, and lifted herself up on tiptoe to kiss Alex deep and long. "You're not going to ignore me this time," she said with a twinkle in her eye. Then she laughed. "Don't worry, Alex, I'm not ready for marriage either!"

At that moment, he heard Dybo's call for help.

※

Terrified out of his wits, Lars Watson struggled to gain control of his fall. He knew his suit had a chute. He had intended to jump anyway as soon as he spread the gel, but that obviously was not going to happen. He seemed unable to get control of his fall, and after several minutes of tumbling, he was completely disoriented. He had no sensation of weight or falling. The world about him seemed to be spinning and lurching in a way that began to make him sick. Then he felt a slight tug, and he began to stabilize. He knew he was way too high yet for chute deployment. He turned and looked over his shoulder, to see two pieces of chute trailing behind him, acting like a stabilizer.

With a great shock, Watson realized that he was looking at the pieces of his chute. Apparently, Dybo had cut the chute when he lunged at him with his utility knife. It took another several seconds for reality to sink in.

Then Watson began to scream…and he screamed…and screamed…and screamed…

PYONGYANG– DPRK (NORTH KOREA)

General Jon Yong-nam sat at his desk contemplating the information he had just received over his encrypted Link. His *EI* people had been taken by the feds, the fool Watson was dead, and Slingshot was up and running. His first action was to sever and completely erase any residual connections between himself and Radler's outfit. Next, he moved several million dollars from the offshore numbered account assigned to *EI* into a numbered account that he could access whenever he wished. Then he closed the *EI* account and wiped out any record of its existence.

When the General was certain that his actions had gone into effect, he punched a secret code into his link. Moments later, an image coalesced over his desk, an image of a man dressed in dark robes, turbaned, with graying hair and a long, full beard.

"Your Grace," the General spoke in English.

The somber image of the newly self-appointed Caliph nodded, acknowledging the General's presence.

"Is the Persian Caliphate ready to discuss the terms for delivery of five hundred intercontinental ballistic missiles?" General Jon Yong-nam asked quietly.

AFTERWORD

NEAR BAKER ISLAND—SUBMERGED ON *WAMPUS*

Margo had just watched Klaus leave to handle a problem with the Western Complex OTEC system. She and he had already made plans with Mabel Fitzwinters to join a second LLI launch loop in the early planning stages. It was slated for construction along the equator in the mid-Atlantic, on floating islands unlike anything ever before attempted. Klaus was to be the engineer in charge, and she had the underwater construction, as with Slingshot. Before they left, Margo insisted on one more dive to visit Amelia at her resting place on the ocean floor off Baker Island.

Margo dropped through the water column in *Wampus*, followed by George and his pals as deep as they could dive. Then she continued by herself, homing on the beacon she had left on the wreck. She reached the bottom and established a hover about twenty meters above the seafloor. She approached the *Electra* with infinite care, so as not to raise any silt. One last time she looked through the shattered windscreen at the physical remains of her heroine, and let a tear trickle down her cheek. *Farewell, my love,* she whispered, *farewell!*

NEAR BAKER ISLAND—ABOARD *RV AMELIA E*

Sometime later, *Wampus* was back in the cradle under the A-frame on *RV Amelia E*. Jeff Carver, the new skipper, manned the controls. Pearl Wells, now the lead diver at the Western Complex, tied down the submersible. Suddenly, Pearl pointed to George and his pals cavorting about a kilometer off the stern. Margo signaled to Carver, who gunned the research vessel, and several minutes later, they drifted alongside George and his friends who were pushing something toward the ship.

"Do you see what I see, Girl?" Pearl asked in astonishment.

Carver stepped out of the pilot house. "That be some food fer thinkin' bout," he said, a wide smile splitting his black Jamaican face.

Margo could hardly believe her eyes. Floating on the surface was a space suit—a space suit with intact helmet, with a body inside. She grabbed a pike and pushed at the suit. It was limp, and bent in ways that a human could not possibly bend. She looked more closely at the bearded face inside the helmet, and realized with a start that it was Neil Lansing—Lars Watson as everyone now knew. He appeared to have hit the water nearly horizontally, and broken every bone in his body. Margo realized that he probably was fully conscious until the very last second. She shuddered as she considered the likelihood.

Using the pike, she pulled Watson's suited body to the divers' ramp at the vessel's stern. "Hold onto him, Pearl," she said as she reached down and twisted the helmet halfway to the left.

A putrid odor wafted up from the collar as she removed the helmet. She held her breath, and indicated to Pearl to push the body beneath the waves. In a moment, Watson's remains disappeared below the surface. Carver made a log entry describing what had happened, and Margo handed him the helmet.

He got the *Amelia E* underway, and headed back to Baker, with George and friends cavorting in the wake.

AMELIA EARHART SKYPORT

The next day found Margo by herself, standing on the main deck of *Amelia Earhart Skyport*. She had requested, and received, permission to ascend entirely by herself. She unwrapped a package she had brought with her—a holograph she had taken the day before of Amelia Earhart's *Electra* sitting upright in magnificent isolation on the ocean floor. She mounted it to the bulkhead and stepped back to look at her handiwork.

"Farewell, my love," she whispered quietly, and then added almost as an afterthought, "*Ad astra!*"

✳ ✳ ✳

Please post a Review for
Slingshot
on
AMAZON.COM and GOODREADS.COM

I really appreciate you posting a review on Amazon and Goodreads. Posting to Amazon.com is intuitive. To post a review on Goodreads.com, go to their website, and become a member if you are not already one. Search for *Slingshot*, and click on the "Want to read" button under the image of *Slingshot*. Indicate that you have read *Slingshot* and then you will be able to post a review. Thank you very much for going through this effort!

Excerpt from the First Chapter of

THE STARCHILD COMPACT
By
Robert G. Williscroft

CASSINI II IN THE ASTEROID BELT

Saeed Esmail prostrated himself toward Earth, nearly 400 million kilometers back in the direction of the Sun. He felt his stomach heave, and vomited blood on his prayer mat, and wondered aloud why Allah had abandoned him. At that moment, he was hit with massive weight, several gees at least, and a twisting, wrenching, totally disorienting surge that made no mental or physical sense. In his weakened state, all Saeed could do was let his body be tossed from wall to wall inside his tent, and hope that he would not tear the airtight fabric. He heard somebody screaming, and then his stomach heaved again, and bloody vomit filled the space around him, flying this way and that, finally collecting on the tent walls. The lights went out, and someone still was screaming, but as the wild gyrations began to settle into a repeating pattern, Saeed realized that he was the one screaming…and he couldn't stop. He reached for his head, pulling out fistfuls of hair…and he screamed again. He retched, but his stomach was empty, and only a little bit of blood mixed with spittle left his mouth, flying at an odd angle to the tent wall…and he screamed, but quieter now, and screamed some more, but quieter still, until his screams morphed into a frightened whimper as he curled into a tight ball on his prayer mat.

L-4—MIRS COMPLEX, FOUR WEEKS EARLIER

A subdued bong captured Saeed's attention. A comforting female voice announced, "In five minutes, we will pitch over and commence our arrival burn at El-four. Please make sure you are securely strapped into your seat, and that you have stowed any loose items you might have been using during the transit. Remain securely fastened in your seat until the arrival announcement tells you it is safe to unbuckle and move about."

Saeed checked his harness, and curiously looked out the port. He saw nothing but stars, more stars than he had ever seen, and off to the rear, the beautiful blue marble that the earth had become—*praise be to Allah*. Then the star field began to rotate, accompanied by a slightly higher pitch from the gyros that penetrated Saeed's conscious perception. The blue marble moved with the star-studded sky until it was positioned above the capsule's port bow. While this happened, Saeed felt no movement. His only sense was that the sky had rotated, as if Allah had reached out and rotated the heavenly backdrop with His mighty hand. Weight returned with a popping hiss as the kick thruster ignited for a few seconds burn. As his weight vanished again, the gyros whined, and the sky began to move from right to left. In short order, Saeed could see the Moon through the ports on the other side of the capsule. It appeared no larger than it did from the Earth, but the left side was one that Saeed had only seen before in holographs. He could not see the Mirs Complex, although he knew it had to lie off the starboard quarter. Weight returned again for about a minute as the restartable kick thruster slowed their velocity to match the orbital velocity of the Russian Federation-built Mirs Complex as it circled the Earth in the Moon's orbit, 385,000 kilometers ahead of the Moon.

Several clanks and surges later, Saeed felt his normal weight gradually return as the capsule nestled into its berth in the capsule arrival bay of the main Mirs Ring, and picked up its rotational speed.

Bong. "Welcome to the Mirs Ring," a bright female voice announced. "It is now safe for you to unstrap and move about. You may disembark to the left side of the capsule. Lavatory facilities are located immediately to the left of the passageway. Your personal belongings will be available in fifteen minutes at the baggage handling dock down the passageway to the right. We know you have choices when traveling off-planet. We thank you for using Slingshot, and hope you had a pleasant trip, and that you will think of us the next time you leave Planet Earth."

Saeed stepped out of the capsule and hurried to the men's room. Although the passengers had been warned about not drinking before the flight, and all the passengers had been issued absorbent diapers an hour before leaving Baker just in case, Saeed, as a faithful Muslim, abhorred fouling himself, and had held off, *by the grace of Allah*, until arrival.

While awaiting the baggage, Saeed checked the construction schedule for *Cassini II*, and then perused the poster-size diagram of the spaceship. *Cassini II* was a sixty-six-meter long twelve-meter wide cylinder, divided into three modules—a twenty-meter long crew module, called the Pullman, a twenty-three-meter long equipment module, called the Box, and the twenty-three-meter long power module and engine cluster, called the Caboose. The large Iapetus-bound spaceship had been constructed entirely at Mirs, about a hundred kilometers away on the opposite side of the main L-4 complex. All three modules had been built in place.

✷

Over the next several days, Saeed mingled with the *Cassini II* provisioning crew that verified the final loadout of the Box and the provisions stored in the Pullman. Another, more technical crew completed the final installation and testing of the gas core reactor and the advanced VASIMR engines that would drive *Cassini II* to Saturn in record time.

On the final day, before the flight crew arrival, the transport tug that ferried the provisioning crew to and from the massive spaceship experienced a catastrophic seal failure where the tug attached to the Box. The entire crew was suited up except, apparently, one Saeed Esmail, the newest provisioning crew member. Searchers found bloody pieces of his suit and a few helmet shards on a trajectory that would ultimately have taken them to the Moon. They never could quite figure out what had actually happened to Saeed, but it was obvious that he had somehow managed to shatter his nearly unbreakable helmet, and rip himself and his tough suit to shreds as he depressurized. The conclusion was that an untracked small meteor, two or three millimeters in size, had gotten him, and somehow maybe even caused the catastrophic depressurization of the tug. Saeed Esmail was not the first casualty on the project, although the consensus was that he might have been the last.

✷

After ejecting the bloody suit pieces and helmet shards from a trash lock in the outer bulkhead of the Box, Saeed worked his way into the hiding place that he had created during the loadout wedged against the outer wall at right angles to both lower level accesses. It was an airtight polymer tent of just over five cubic meters, with its own oxygen supply and scrubber. It would keep him alive during the transit to Iapetus. He had the freeze-dried food, water from the emergency supply, and he could dump waste out the waste lock. His Link, with its collection of holofilms, books, and the *Qur'an*, would keep his mind occupied for the projected four-month trip. He examined the four burst transmitters that had been included in his life pack. About the size of a softball, each was designed to be ejected through the waste lock, orient itself with the ship to its rear, extend a gossamer parabolic antenna, and do a circular search for Earth, using a very limited supply of compressed gas. Then, using a high-density charge, the device would transmit a series of encrypted bursts until the charge was consumed. Saeed was to deploy the first at the tether extension, the second following the Jupiter boost, the third when they arrived near Saturn, and the fourth was for whatever circumstance warranted a special transmission.

In his hideaway, Saeed prostrated himself facing Earth, he hoped, and recited his prayers, adding a personal thanks to Allah for keeping him safe thus far, and on line to accomplish His holy mission.

※

Jon Stock stepped out of the launch loop capsule at Mirs Ring and made a beeline for the men's room. "Those capsules need a latrine," he muttered to himself as he splashed water on his face. Steely blue eyes stared out at him from the mirror. His hair was gray and cropped short above a craggy, clean-shaven face that testified to his fifty years. A lean, muscled 183-centimeter frame belied those same years. He wore the uniform of a U.S. Navy Captain, his left chest bedecked with ribbons. One stood out top center, jet black, framed in silver, with a golden image of Mars attached to the center—the Mars Expeditionary Medal. Jon was the second in command on that first expedition to the Red Planet. When Commander Evans was killed in a freak accident on the surface, he assumed command, saved the mission, and brought the crew back. Now he commanded the international crew of *Cassini II* on an expedition to Iapetus. They would travel five times further than any human had ever gone before. And what awaited them at their destination might very well change human history forever.

Iapetus… Jon reviewed what he knew about Saturn's iconic moon. In 2004, the *Cassini-Huygens* spacecraft flew by Iapetus. Iapetus proved to be unlike any other moon. The surface seemed to display an intersecting grid of geodesic sections, something not normally found in nature. A narrow mountainous wall extended around Iapetus at the equator, so that the moon looked something like a walnut. Iapetus' density was far too low for a moon that appeared solid, but if Iapetus were substantially hollow, then the numbers worked out just about right. Several of the "geodesic sections" appeared to have collapsed inward, revealing what could be interpreted as complex structures underneath the surface layer. A tall, very narrow structure extended from the surface at one point, like a towering spike a kilometer high. Like the "geodesic structure," this spike had no "natural" explanation.

In September 2007, the *Cassini-Huygens* spacecraft made another relatively close transit of Iapetus following the equatorial wall, revealing that the wall consisted of a series of mountains up to twenty kilometers high, following each other in series, none side-by-side. It also supplied further details on a series of equally spaced craters parallel to the equatorial wall and halfway between the wall and the North Pole.

Iapetus had remained a mystery. It was very difficult to imagine that all the things discovered by *Cassini-Huygens* were natural. The implications of the discoveries being artificial were staggering. As more and more information was gathered by space telescopes in orbit around Earth, on the Moon, and at the Mirs Complex at L-4, the possibility that Iapetus could have an artificial origin became quite real. The initial concept for a human investigation of Iapetus had been put forward by Launch Loop International (*LLI*), the consortium that had built Slingshot as an entirely civilian operation, followed by several other launch loops around the world. While there was lots of pushing and shoving by the governments of the territories where the launch loops were located, in the final analysis, most people considered a launch loop as something akin to an airline company, and in the end, most of the loops were left in civilian hands, although governments exercised whatever control they wished.

Iapetus, however, was seen by the world's major players as a potential prize like none other. If Iapetus turned out to be an artifact, eloquent spokespersons from various governments argued, then it belonged to all the people, not just to the greedy corporations that found it. This argument fell on sympathetic ears of a world population that had grown used to being told what to do by *benevolent* governments. When *LLI* partnered with their former rival, Galaxy Ventures, to form Iapetus Quest, they found themselves faced with an unusually consolidated array of governments united in their opposition to a privately funded and operated Iapetus operation. The United States, in its still dominant position on the world stage, muscled itself into the leadership slot in the newly recast government-owned and operated Iapetus Quest. The international debate had raged on how to structure the crew of *Cassini II*. Many had argued for a civilian crew, structured however they wanted. Eventually, by negotiated treaty, arm twisting, back-room dealing, and even outright bribery and coercion, an international crew was assembled that represented the interests of the participating nations.

You have just been reading from Chapter One of The Starchild Compact, *the second volume in Robert G. Williscroft's exciting Science Fiction series,* The Starchild Trilogy. *Buy it wherever books are sold.*

WORDS OF PRAISE FOR *THE STARCHILD COMPACT*

In the not-too-distant future, a spacecraft heads toward Saturn's moon Iapetus to investigate whether it is an artifact, while a terrorist stows away on board hoping to destroy the science that contravenes the tenets of his religion. All this builds up the tension and suspense in this fascinating science fiction novel. Each part of this book solves and unfolds another mystery, making the book incredibly hard to put down. The research and science are impeccable. I marveled at Williscroft's imagination in conjuring up this story. I highly recommend this book!

– Marc Weitz, Past President
The Adventurers' Club of Los Angeles

Hard sci-fi reminiscent of Arthur C. Clarke or James P. Hogan, with a geopolitical twist worthy of Tom Clancy or Clive Cussler.
—Alastair Mayer
Author of the *T-Space Series*

The Starchild Compact is a compelling read from the first page. Robert has written a fantastically engrossing space mystery that takes place in our own backyard. This book brought me moments of wonder that I had experienced when I originally read Clarke's *Rendezvous with Rama*. This does what science fiction is supposed to do: capture our attention, speculate about the wild possibilities, and take us just beyond our previous imaginings. But this book is not all spectacle. Robert tackles some of the more personal issues of space travel that often go overlooked, with a particular eye toward the role of religion in that exploration. It is a masterful hand that can manage the personal and cultural response to the wonders of space and still present those wonders as pure delight. Robert has done that in *The Starchild Compact*. From the beginning to the end, this is a must read.
— Jason D. Batt
100 Year Starship
Author of *The Tales of Dreamside Series*

Robert Williscroft once again delivers. Readers unfamiliar with Williscroft will be amazed at the depth of his characters and his meticulous science and engineering. *The Starchild Compact* is a remarkable story of politics, intrigue, science, engineering, and derring-do, driven by imaginative speculation. Nine exceptional men and women from divergent backgrounds undertake a voyage of discovery. Against a backdrop extrapolated from today's headlines, they struggle to accommodate their differences, while meeting the challenges a hostile universe throws at them as they journey to Saturn's moon Iapetus, all-the-while dealing with a Jihadist stowaway from the Persian Caliphate, a nuclear-armed world-power in this near future. They determine that Iapetus is an artifact, and discover its origins. They meet the Founders—direct descendants of the Iapetus architects. Who and what the Founders are profoundly affect not just the voyagers and the Jihadist, but all the peoples of Earth.

The author's fans, as well as new readers who crave anthropological authenticity and honest-to-Heinlein Hard Science Fiction, will be thoroughly delighted with *The Starchild Compact*.

– Martin Bloom, President
The Adventurers' Club of Los Angeles

In *The Starchild Compact* Robert Williscroft has said in print what a lot of people (myself included) would like to do about present day threats to our democracy and way of life, but don't have the means or cojones to do it. He also courageously extrapolates tomorrow's mores and the religious direction our society is taking. Williscroft tackles these germane and "heavy" issues while crafting a fascinating novel that is hard to put down. I have to admit that I was moved to tears, because I could not be with the space travelers to come back and see Earth's future. I'm looking forward to both the prequel and sequel.

— Myron R. Lewis, Co-author with Ben Bova of several
SF stories including parts of *The Dueling Machine*

Intrigue and danger blended with today's societal problems carry the reader on an unexpected journey. An internationally diverse spaceship crew comes together to face their differences and potential dangers on a voyage to Saturn's moon Iapetus, which they suspect may be an artifact. Their individual quirks and cultural traditions come face-to-face with the reality of a new paradigm with global repercussions when the crew discovers irrefutable evidence that the builders of Iapetus still have a presence in the Solar System. Highly recommended!

– Matthew Severe
Author of *The Lariat Thief*

Hyperchess™

Hyperchess is a chess player's game. The word Hyperchess is a contraction of the two words hyperspace and chess. Hyperspace, as any reader of science fiction knows, is a characteristic of space often employed in science fiction to effect faster-than-light travel from one point to another, sometimes viewed as another dimension or a parallel universe. Hyperchess incorporates this other-dimensional concept into the game of chess. In chess, all moves are on the chessboard in the physical universe occupied by the pieces. In Hyperchess, however, every move is either through "hyperspace" or into the "parallel dimension." This dramatically changes how the game is played, and produces an entirely set of new strategies.

Hyperchess Rules

Equipment

No special equipment is needed to play Hyperchess. The game can be set up two ways. The more expansive, visual approach uses one set of pieces and two stacked boards oriented the same way, stacked with sufficient room between the boards to move the pieces. The game can be played on one board, however, by using two distinctly different sets of pieces (ideally shaped the same, but with contrasting colors, e.g. black and red opposes white and blue), and four clear glass tumblers sufficiently tall to be placed upside down over any of the pieces[1].

Notation and Piece Movement

POSITIONAL DESIGNATION. To understand positional designation, visualize the two stacked boards. Pieces on the upper board are U-pieces. Pieces on the lower board are L-pieces. For single-board play, pieces belonging to one set (e.g. the black and white set) are U-pieces, and pieces belonging to the other set (i.e. red and blue) are L-pieces. Every piece carries its traditional notation followed by "[u]" or "[l]" (e.g. P[u], meaning an upper pawn, or KB[l], meaning a lower king-bishop).

---

1   Any method of distinguishing between U and L-pieces works. Clear plastic thimbles placed on a piece can designate it as a U-piece, or even a bit of modeling clay. Anything at all that distinguishes U from L-pieces and does not interfere with the game will work.

Starting Positions. "White" pieces start in their traditional positions as U-pieces. "Black" pieces start in their traditional positions as L-pieces.

The Fade. During the course of play, any piece except the knight can fade from one board to the other by moving to the same square on the other board, so that its positional designation changes. Such a move is notated by an italicized lower-case letter "*f*" between the old and the new positional designation (e.g. P[u] *f* [l] indicates P[u] fades to become P[l]).

The Transit. The traditional chess move for pieces other than a knight is replaced by a transit. A transit is defined as a fade followed by the traditional move for the piece followed by another fade. When transiting, a piece fades to the other board, makes whatever authorized move it can, and fades back to the original board—in effect it transits through hyperspace. If the square on which a piece lands is occupied by an enemy piece, that piece is captured. A transit is notated exactly like traditional chess, except that each piece always carries its positional designation (e.g. P[u] to Q4[u] or P[l] to P3[l]).

Note that when playing on a single board, a transiting piece can pass through any piece with the same positional designation, and it can only capture a piece with the same positional designation.

The Jump. The knight's move combines the fade and transit into a single move—the jump. The knight makes its traditional chess move, but fades either at the beginning or at the end of the move (but fades only once), so it always ends up on the opposite board. As in traditional chess, a knight can pass through any piece, but its destination square must be unoccupied or occupied by an enemy piece that it captures. A jump is notated exactly like traditional chess, except that the knight always carries its positional designation (e.g. Kn[u] to R3[l] or Kn[l] to QB3[u]).

The Cage. A consequence of jumping a knight is that it can end up in the square above or below any piece from either side. When this happens, that other piece is caged. A caged piece other than a knight cannot be moved. When playing on a single board, a cage is indicated by placing the tumbler upside down over the caged piece, and placing the caging knight on top of the tumbler. A caged piece other than a knight cannot move until either the caging knight moves or is captured. A caged piece can only be captured by the other knight.

CAPTURING. Capturing is the same as in traditional chess. An opposing piece is captured when it occupies the ending destination of a piece, after a transit or a jump (but not a fade).

CHECK, MATE & STALEMATE. These are the same as in traditional chess. Note, however, that a king can fade out of a check, unless the destination square is guarded. A guarded square is a potential terminal position of any opposing piece that is not caged. A king can be caged by either side. A caged king checked by an opposing knight is mated unless the checking knight can be captured. It can only be mated by the other knight, and is held in the cage until the caging knight is either captured (by an opposing knight only) or moves.

Special Situations

CASTLING. As in traditional chess, castling pieces may not move through guarded squares.

EN PASSANT. En passant is as in traditional chess, but happens following the transit rules.

KNIGHT COLOR-DEPENDENCE. A knight always jumps to the opposite board. As a consequence, it is restricted to the square-color of its origin-square on its origin-board and the opposite square-color on the other board. It can only cage pieces of the opposite square-color on the origin-board and the same square-color on the other board (e.g. the white QKn is initially a U-piece on a white square; it can cage only black-square U-pieces and white-square L-pieces).

BISHOP COLOR-DEPENDENCE. A white-square bishop is restricted to the white squares on both levels, and a black-square bishop is restricted to the black squares on both levels as in traditional chess. As a consequence, no matter what level a bishop occupies, it retains its square-color dependence.

DISCOVERED CHECK. As in traditional chess, a discovered check happens when moving one piece exposes the opposite king to attack from another piece. In Hyperchess, when the square occupied by a king is guarded by a caged piece, the king is not in check, but when the caging knight moves (and thus removes the cage), the king is then in check by the released piece.

OPENING PLAY. Unlike traditional chess, any major piece can be transited into play without first moving a pawn. One possibility is to transit a major piece forward on the first move, and fade it to a guarded position on the next move. In this way it threatens the opposing second line. To avoid capture, the opposing side will be forced to fade a major piece or fade a pawn into the attacking transit lane. No opening strategies have been developed yet, so the field is wide open.

Communicate with the inventor.

If you enjoy Hyperchess, and especially if you discover strong strategies, please email the inventor with your information to *hyperchess@argee.net*. In time, the inventor will distribute a periodic email newsletter containing these developing strategies. Join the newsletter list by sending an email to the above address.

About the Author

At the Adventurers' Club of Los Angeles

Dr. Robert G. Williscroft is a retired submarine officer, deep-sea and saturation diver, scientist, author, and a lifelong adventurer. He spent 22 months underwater, a year in the equatorial Pacific, three years in the Arctic ice pack, and a year at the Geographic South Pole. He holds degrees in Marine Physics and Meteorology and a doctorate for developing a system to protect SCUBA divers in contaminated water. A prolific author of both non-fiction, Cold War thrillers, and hard science fiction, he lives in Centennial, Colorado.

Dr. Williscroft is a member of Colorado Author's League, Independent Association of Science Fiction & Fantasy Authors, Science Fiction Writers of America, Libertarian Futurist Society, Los Angeles Adventurers' Club, Mensa, Military Officer's Association, American Legion, and the NRA, and now spends most of his time writing his next book, speaking to various regional groups, and hanging out with the girl of his dreams, Jill, and her two cats.

Other books by this author

Please visit your favorite eBook retailer to discover other eBooks by Robert G. Williscroft and your favorite online bookseller for their paper versions:

Current events:
The Chicken Little Agenda—Debunking "Experts'" Lies

Children's books:
The Starman Jones Series:
  Starman Jones: A Relativity Birthday Present
  Starman Jones Goes to the Dogs (2022)

Short Stories:
The Daedalus Files:
  Daedalus
  Daedalus—LEO
  Daedalus—Squad
  Daedalus—Combat

Novels:
The Daedalus Files
Mac McDowell Missions:
  Operation Ivy Bells
  Operation Ice Breaker
  Operation Arctic Sting
  Operation White Out (2022)
The Starchild Trilogy:
  Slingshot
  The Starchild Compact
  The Iapetus Federation
The Oort Chronicles:
  Icicle—A Tensor Matrix
  Federation—To the Stars
  Andromeda—A Rising Tide (2022)

Connect with Robert G. Williscroft

I really appreciate you reading my book! Here are my social media coordinates:

Facebook: *https://www.facebook.com/robert.williscroft*
Parler: *https://parler.com/#/user/RGWilliscroft*
Gab: *https://gab.com/RWilliscroft*
Twitter: *https://twitter.com/RGWilliscroft*
LinkedIn: *https://www.linkedin.com/in/argee/*
Instagram: *https://www.instagram.com/rwilliscroft/*
Pinterest: *https://www.pinterest.com/RGWilliscroft*
YouTube: *https://www.youtube.com/user/RWilliscroft*
Book Website: *https://RobertWilliscroft.com*
Personal Website: *https://argee.net*
Blog: *https://thrawnrickle.com/*
Publisher: *https://freshinkgroup.com/author/robertwilliscroft/*
Amazon Author Page: *https://www.amazon.com/Robert-G-Williscroft/e/B001JP52AS*

Glossary

Abyssal plain—An underwater plain on the deep ocean floor, usually found at depths between 3,000 and 6,000 m.

Aku Aku—Mothership for the *Alvin*.

Alim—An over-the-counter medication available in the time-frame of the story that eliminates alcohol from the body.

Alvin - Mini-sub capable of depths in excess of 5,000 meters.

Amaranth—A Barquentine that foundered on the southern shores of Jarvis in 1913.

Amphib—A plane that is capable of landing on either water or land.

Apex—That point on the Eastern and Western Complexes where the Downslope enters the water and the Upslope exists the water. The narrow part of the complex where the facilities are located.

Aramid fiber—A light-weight synthetic fiber that has a tensile strength stronger than steel.

Baker Compound—The Slingshot facility on Baker Island.

Barquentine—A sailing vessel with three or more masts, with a square rigged foremast and all other masts fore-and-aft rigged.

Beanstalk—A space elevator as conceived by Jerome Pearson.

Bitter end—The end of a rope, line, or cable.

Bone conduction headset—A headset that conducts sound to the inner ear through the bones of the skull. In use, a rubber-covered small speaker is strapped against one of the dome-shaped bone protrusions behind the ear and the sound, which can be surprisingly clear and crisp, seems to come from inside the user's head.

Carbon nanotubes—Tubular cylinders of carbon atoms that have extraordinary mechanical, electrical, thermal, optical, and chemical properties.

Coxswain—The driver of a boat.

Deflector—A series of permanent and electro-magnets that bend the path of the rapidly moving iron ribbon.

Der Schimmelreiter—From a novella of the same name by German writer Theodor Storm. Loosely translated, it means The Rider on the Gray Horse, and refers to this rider cruising a dyke.

DIW—Dead in the water.
Eastern Complex—The eastern in-water facility that houses the OTEC generator, the linear drivers, and the controlling facilities, along with housing.
Ecobabble—Using the technical language of ecology to make the user seem ecologically aware.
EFCom—Electrostatic Field Communication System.
ETA—Estimated time of arrival.
Explosive cutter—An explosive device that cuts a cable on demand.
Fake, fake down—To lay a rope or line in a series of overlapping back-and-forth folds to allow it to pay out freely.
Fantail—The rear open deck of a ship or boat.
Ferro-cement—A building material made of thin cement slabs reinforced with steel mesh; the cement is often whipped to a foam to give it lightness.
Frictionless motion deflectors—The deflectors constructed of permanent and electro-magnets that bend the path of the rapidly moving iron ribbon.
FHWA—Federal Highway Administration.
Gunwale—The top edge of the side of a boat.
Howland Island—A coral island in the equatorial Pacific about sixty-five kilometers north of Baker Island. It was the destination of Amelia Earhart when she disappeared.
Hyperchess—A variation of chess played by the protagonists (see Hyperchess instructions herein).
Jarvis Compound—The Slingshot facility on Jarvis Island.
Keith Lofstrom—Inventor of the Launch Loop.
Kevlar—A brand name for aramid fiber, especially cloth made from aramid fiber.
Kick thruster—A small, reigniteable solid-state rocket attached to a capsule, used for vector changes after release from the rail, or to slow down a capsule used to transit from Baker to Jarvis.
Klick—Slang word for kilometer.
KP—Portable one-man recompression chamber made of reinforced Kevlar and other materials.
Lanthanide series—That portion of the Periodic Table from Lanthanide to Lutetium.
Launch facility—General term referring to a facility where something is launched, like a sub, boat, capsule, etc.

Launch Loop—The subject of this novel. A means for getting into space without using rockets.

Launch pouch—Attaches to the capsule underside, enabling magnetic acceleration of the capsule by the rail.

Linear accelerator—A magnetically induced form of acceleration where an object is accelerated by a rapidly moving magnetic field.

MAST suit—Medical Anti-Shock Trousers—medical devices used to treat severe blood loss on an emergency basis while transporting the patient to a medical facility.

Nikumaroro Island—A remote, elongated, triangular coral atoll 550 kilometers south of Howland Island. Some members of TIGHAR believe Earhart ended up here.

Orbital Ring—An Earth-circling cable with skytower stations, conceived by Paul Birch.

OTEC—A power generation system that uses the temperature difference between cooler deep and warmer surface ocean waters to produce electricity.

Outboard—A boat motor that is attached to the stern of a boat, as opposed to an internal engine.

Paul Birch—British inventor of the Orbital Ring, which reduces to a Launch Loop when simplified.

Pigtail—A loose end of something, typically a rope or line.

Port of Lae—The largest Port in Papua New Guinea.

Radionav gear—A radio direction finder that assists in navigation at sea and in the air.

Rail—Common term for the portion of the launch loop between the skyports.

Rebreather—A diving air source that reuses the nitrogen in the air, while scrubbing out the carbon-dioxide and replacing it with oxygen.

Recompression chamber—A steel chamber used to recompress divers suffering from decompression sickness (the Bends).

Ribbon—Common term for the soft-iron tube that is the heart of the launch loop.

Sikorsky Skycrane—A twin-engine heavy-lift helicopter.

Skydome—The sky as it appears on the open ocean, from a small island in otherwise open ocean, or out on an open prairie.

Skyport—The structure at the top of the skytower.
Skyrail—An alternative name for a Space Elevator or Launch Loop.
Skytower—The elevator-like set of cables that extends from the Skyport to the island below.
Socket—The attachment point on the island for the skytower.
Space elevator—A cable hung from a geostationary satellite to a stationary place on Earth. Elevator cars run up and down this cable allowing people and cargo access to space. There is currently no material sufficiently strong to carry its own weight in this configuration.
SSRS—Spread-spectrum radio system.
Surface-effect boat—A boat that employs captured air held between skirts or thin hulls to hold a boat's main hull out of the water.
Suspensor cable—A cable to which the skytower cable and double-lift cables are attached. It carries the weight of all the cables.
T-anchors—Anchors with a form of barb that holds the anchor body to the sea floor.
Tensioner—A cable attached to the rail or downslopes and the ocean bottom, with a dynamic device that increases or decreases the tension as necessary to maintain Launch Loop stability.
Terminus—The triple point below the skyport where the downslope/upslope, skytower, and rail come together.
TIGHAR—The International Group for Historic Aircraft Recovery.
Toot-ta-doo—The cardboard cylinder that toilet paper or paper towels come on. Name derived from the inability to resist picking up an empty one and making a Kazoo like toot-ta-doo noise through it.
Transom—The flat surface forming the stern of a boat.
UV—ultraviolet.
Western Complex—The western in-water facility that houses the OTEC generator, the linear drivers, and the controlling facilities, along with housing.
Whitehall—A metonym for the British government.
Windlass—A type of winch used to move heavy things.

Fresh Ink Group
Independent Multi-media Publisher
Fresh Ink Group / Push Pull Press / Voice of Indie

☙

Hardcovers
Softcovers
All Ebook Platforms
Audiobooks
Worldwide Distribution

☙

Indie Author Services
Book Development, Editing, Proofing
Graphic/Cover Design
Video/Trailer Production
Website Creation
Social Media Management
Writing Contests
Writers' Blogs
Podcasts

☙

Authors
Editors
Artists
Experts
Professionals

☙

FreshInkGroup.com
info@FreshInkGroup.com
Twitter: @FreshInkGroup
Facebook.com/FreshInkGroup
LinkedIn: Fresh Ink Group

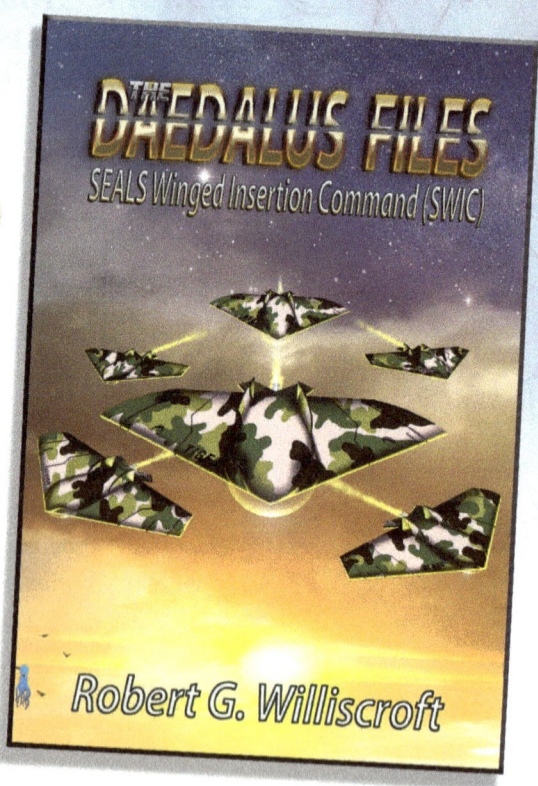

In *Daedalus*, Navy SEAL Derek "Tiger" Baily, irreverent member of the SEALS Winged Insertion Command (SWIC), makes a harrowing first base jump in the experimental *Gryphon-7* hardshell wingsuit from the edge of Space. He test-flies the armor-plated *Gryphon-10* in *Daedalus LEO*, catapulting into space by Slingshot and dropping from a record-obliterating 160 klicks. Testing the enhanced *Gryphon-10 MK 4* in *Daedalus Squad*, Baily's 6-man team believes it is fully prepared for hurtling around the world and staging critical re-entry, but challenged to innovate life-or-death solutions with only seconds to spare, it might not survive intact. Then the presidential front-runner is seized by pirates for ransom in *Daedalus Combat*, and the SWIC team is called to action where it must literally improvise on the fly with everything to lose. Join Tiger Baily through all four adventures in sci-fi master Robert G. Williscroft's *Daedalus* series, now collected for the first time as *The Daedalus Files: SEALS Winged Insertion Command (SWIC)*.

Hardcovers, Softcovers, Ebooks

Fresh Ink Group
www.FreshInkGroup.com

His head frozen at death, engineer and entrepreneur Braxton Thorpe awakens a century later to existence inside an electronic matrix. He discovers the GlobalNet served by ServerSky, and taps into databases and expands to become a Tensor Matrix with access to virtually all human knowledge. He works with Daphne O'Bryan and other researchers, then is joined by an extraordinary non-human partner, the Oort, as he generalizes his existence from Banach Space into a Banach Manifold. Using complex tensors to extend beyond the matrix, he discovers a malevolent threat to the entire Solar System, marauding Aliens with bad intent. Braxton and Daphne, assisted by the Oort, must build a defense against this advanced alien civilization, but can they do this in time, or will it be too late for mankind? **Now in Jacketed Hardcover, Trade Softcover, all Ebook Formats, and Audiobook!**

Fresh Ink Group
FreshInkGroup.com

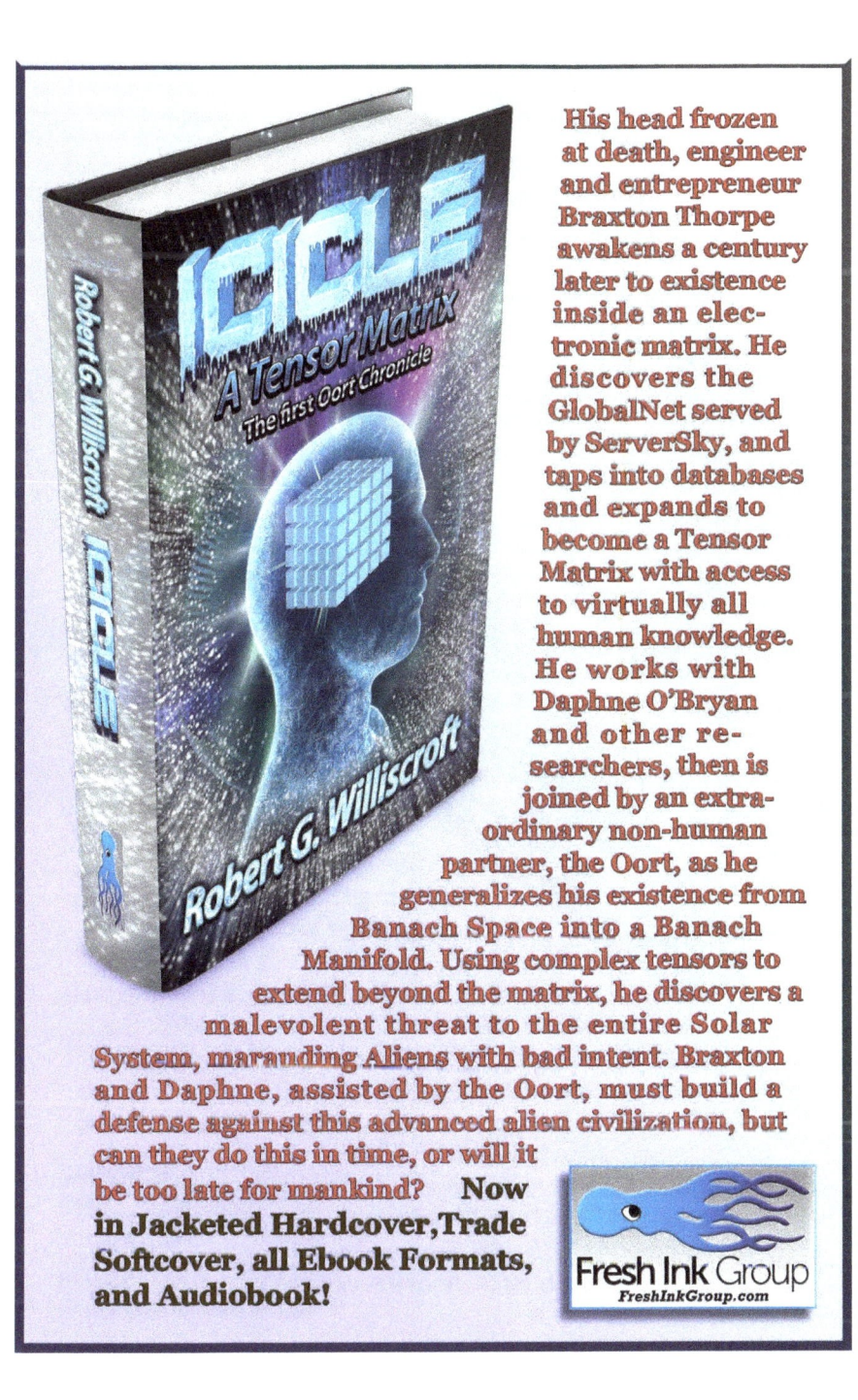

As the Oort Federation becomes a major force in the Solar System, Braxton Thorpe passes the Federation chairmanship to former U.S. President John Butler. Thorpe's group offers humanity virtual immortality, but Isidor Orlov and his Udachny Enterprises oppose their every move. Phoenix terraforms Mars for more living space, but the Mars Reds prove formidable resisters. If the Asterian starship fighter pilots are released, will they align with Phoenix or Udachny, and who will develop the right FTL technology? In this tense space adventure, Thorpe, his team, and Max the tabby cat travel to Proxima Centauri and beyond to the Aster system, 84 lightyears distant. Will Thorpe bring together humans and Asterians in their quest for intergalactic travel? Will long life prove more than mere humans can handle? *Hardcover, Softcover, All Ebook Formats, Audiobook with SFX & Music!*

Fresh Ink Group
FreshInkGroup.com

OPERATION IVY BELLS
A MAC McDOWELL MISSION
Robert G. Williscroft
Foreword by Ed Offley, author of Scorpion Down

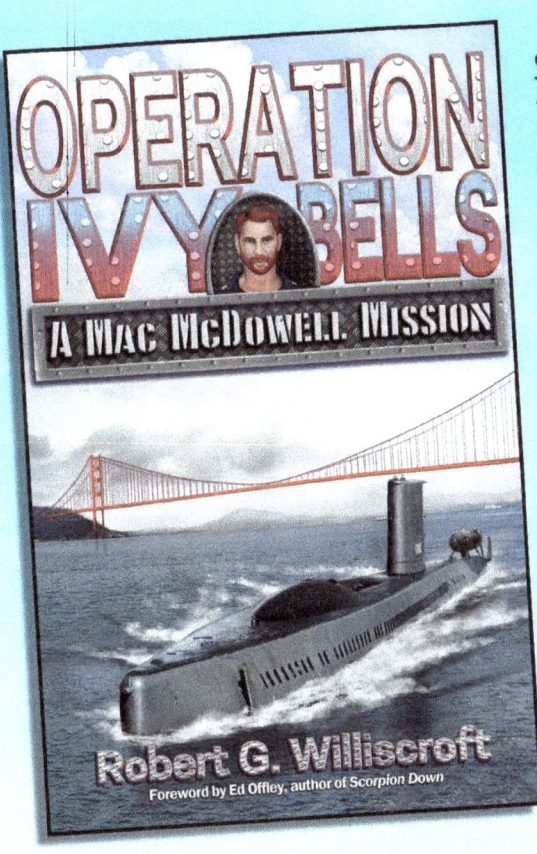

Saturation Dive Team Officer-in-Charge (OIC) Mac McDowell faces his greatest challenge yet, leading the team into a critical Cold War mission. With a security clearance above Top Secret, Mac and his off-the-books deep-water espionage group must gather Russian intel to avert world war. Join nuclear-submariner Mac as he extreme-dives to a thousand feet, battles giant squids, and proves what brave men can achieve under real pressure, the kind that will steal your air and crush the life out of you. *Operation Ivy Bells: A Mac McDowell Mission* updates the popular bestseller by Robert G. Williscroft, a lifelong adventurer who blends his own experiences with real events to craft a military thriller that will take your breath away.

Fresh Ink Group

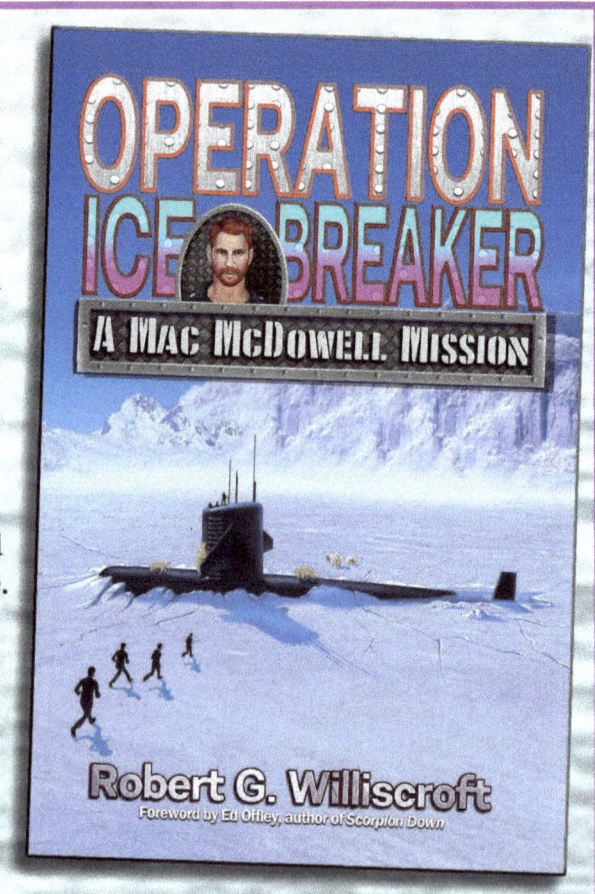

USS Teuthis Saturation Dive Team Officer-in-Charge (OIC) Mac McDowell is leading his submarine team on a mission to lay SOSUS arrays under the ice in the Arctic when they clash with a new Alfa-class highly automated Soviet submarine. Overwhelmed by mechanical problems, the Soviet crew abandons their sub near Pt. Barrow, Alaska. The *Teuthis* skipper launches *DSRV-1 Mystic*, so Mac and his crew can board the empty sub and gather intelligence. The arrival of an even more advanced Soviet sub leads to breathtaking underwater clashes with the specter of war looming. Will the Soviets sink Mac and his crew to their watery graves, or will Teuthis safely return to Alaska where Mac's new love, Kate, anxiously awaits his return?

Hardcover, Softcover, All Ebooks, Audiobook

Fresh Ink Group
FreshInkGroup.com

As *USS Teuthis* Saturation Dive Team Officer-in-Charge (OIC) Mac McDowell leads his submarine team laying SOSUS arrays under the Arctic ice, they capture an abandoned fully automated Alfa-class Soviet sub. Piloting their prize through the ice pack to the U.S. East Coast, they must evade or confront other Soviet subs trying to recover the sub—or sink it. Breathtaking deep-sea clashes erupt, including hand-to-hand combat with Soviet Morskoy Spetsnaz divers under the ice. Too far from *Teuthis* to escape, the Americans are accosted by a 5-ton orca. Will Mac's ship survive long enough to reach friendly waters, or will the men become just another meal for a deadly whale?

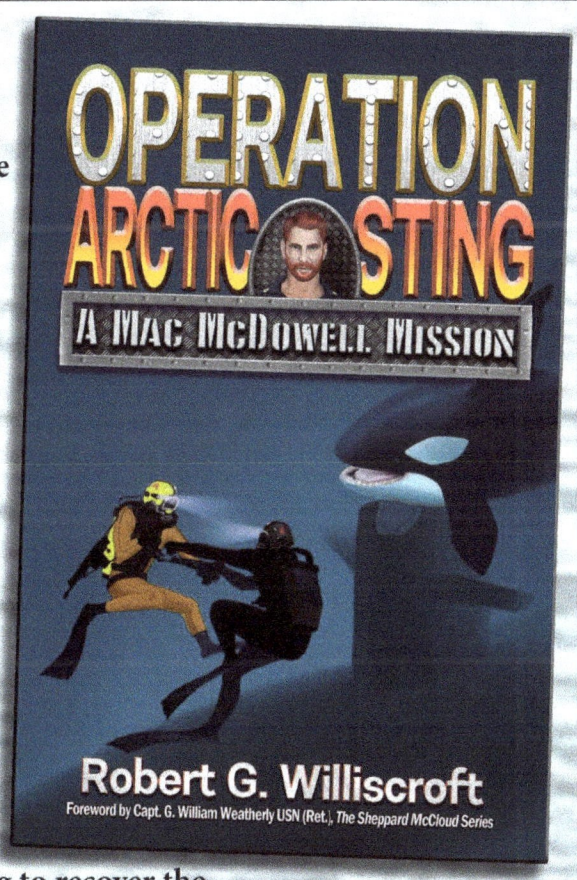

Hardcover, Softcover, All Ebooks, Audiobook

OPERATION ARCTIC STING
A MAC McDOWELL MISSION

Robert G. Williscroft
Foreword by Capt. G. William Weatherly USN (Ret.), *The Sheppard McCloud Series*

Fresh Ink Group
FreshInkGroup.com

Recovering from his Operation Arctic Sting injuries, USS Teuthis Executive Officer Mac McDowell is tasked with laying SOSUS arrays in the southern Atlantic and off Thurston Island, Western Antarctica. Teuthis tangles with Argentine subs in the south Atlantic, then confronts a ChiCom sub off Thurston Island. Mac and his team experience serious setbacks at the hands of the ChiComs while installing a relay transmitter on a nearby mountain peak. Teuthis discovers an underwater oil operation off Thurston Island and is tasked with escorting a Taiwanese sub and underwater tanker under the cover of the largest military marine exercise since World War II: PacEx89. Teuthis is attacked by a Chinese Han-class sub and a previously unknown North Korean AIP sub despite the protection provided by three U.S. fast-attack subs. Will Mac and Teuthis complete their mission, or will they finally meet their watery graves on the Pacific Ocean abyssal plain?

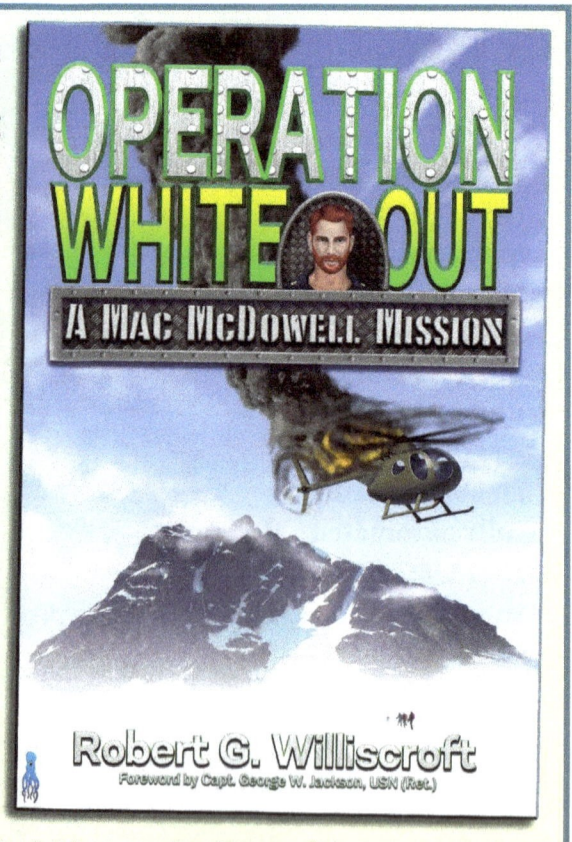

Hardcover, Softcover, Digital, Audiobook

Fresh Ink Group
FreshInkGroup.com

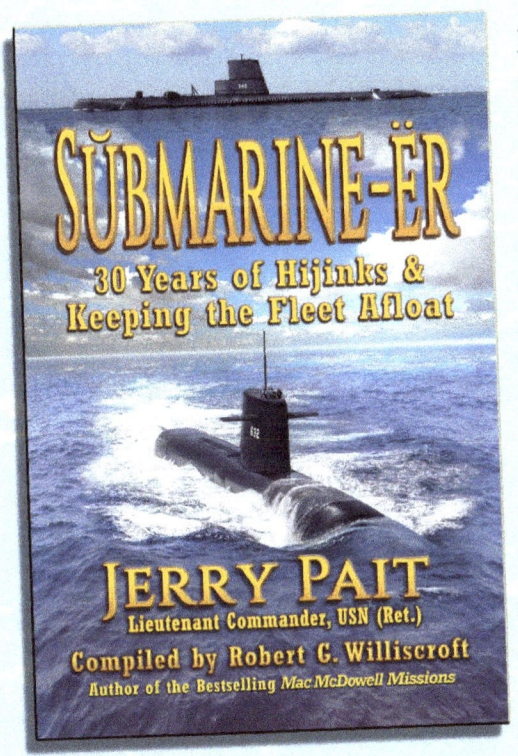

SŬBMARINE-ËR
30 Years of Hijinks & Keeping the Fleet Afloat

JERRY PAIT
Lieutenant Commander, USN (Ret.)
Compiled by Robert G. Williscroft
Author of the Bestselling *Mac McDowell Missions*

Lieutenant Commander Jerry Pait's semi-autobiographical collection of sixty stories recounts his thirty years in and around the U.S. Navy's sub- marine fleet. Ranging from light-hearted to wrenching, all are poignant inside looks at naval operations rarely seen by outsiders. Topics include the real story behind the shuttle Challenger tragedy, risking his own life underwater, discovering a Soviet spy living across the street, surviving when a DELTA Rocket engine ignites, critical missions, and the everyday lives of men and women of the fleet. Dive into Sŭbmarine-ër for hijinks and breathtaking adventure with this poignant memoir by a true American hero.

Fresh Ink Group
FreshInkGroup.com

Hardcover
Softcover
All Ebooks
Worldwide